The Alpaca
Its Naturalization in the British Isles Considered A Benefit To The Farmer

by Wm. Walton

with an introduction by Jackson Chambers

This work contains material that was originally published in 1845.

This publication is within the Public Domain.

*This edition is reprinted for educational purposes
and in accordance with all applicable Federal Laws.*

Introduction Copyright 2017 by Jackson Chambers

Self Reliance Books

Get more historic titles on animal and stock breeding, gardening and old fashioned skills by visiting us at:

http://selfreliancebooks.blogspot.com/

Introduction

I am pleased to present another title in the "Alpaca" series.

The work is in the Public Domain and is re-printed here in accordance with Federal Laws.

As with all reprinted books of this age that are intended to perfectly reproduce the original edition, considerable pains and effort had to be undertaken to correct fading and sometimes outright damage to existing proofs of this title. At times, this task is quite monumental, requiring an almost total "rebuilding" of some pages from digital proofs of multiple copies. Despite this, imperfections still sometimes exist in the final proof and may detract from the visual appearance of the text.

I hope you enjoy reading this book as much as I enjoyed making it available to readers again.

Jackson Chambers

THE ALPACA.

CHAPTER I.

HISTORY AND PROPERTIES OF THE ALPACA.

At the the time when Pizarro and his twelve hardy followers first reached the Peruvian shore, (1525,) among other objects of interest and curiosity, they found the natives in possession of two domestic animals, each peculiar in its kind, and differing from any which they had previously seen in the newly-discovered regions of the west. To the Europeans these animals appeared to be an intermediate species between the camel and the sheep, but partaking more of the properties of the latter; in consequence of which, filled with the ideas of fatherland, they called them *carneros de la tierra*, or country sheep, the designation by which the llama and alpaca were afterwards known to the Spaniards. Having completed his discovery, Pizarro returned home, provided with reports on his success, and descriptions of the varied productions noticed by him in the new region, which he offered to add to the dominion of Castile; and, as tradition states, among his packages were specimens of alpaca wool, together with textures made from it by the natives.* His arrival excited the liveliest sensation among his countrymen; and so much was Charles V. delighted with the prospect of so valuable an acquisition as the one described to him, that he conferred the supreme command of the force solicited upon the discoverer, through whose instrumentality he hoped to see the conquest achieved, at the same time loading him with honours.

Seville was then the court of the Castilian monarchs, as well as the emporium of that commerce which the discovery of the New World had created. It was not only the chief city in Spain, but also the metropolis of an empire comprising the most flourishing countries in Europe. There the most distinguished individuals from Italy, Germany, and the Low Countries, met in union with their Spanish fellow-subjects before the throne of their general master, vying with each other in their ardour to execute his commands. The cleverest Spaniards in every branch of science were congregated on the banks of the Gaudalquivir; in every department of war and politics, no government was better served than that of Spain; and it is only by taking these circumstances into consideration, that we can duly appreciate the talents, and estimate the character, of those individuals selected to accompany or follow Pizarro, for the purpose of preaching the gospel and organizing a government in Peru, and to whom we are indebted for the first accounts of that interesting race of quadrupeds to which the alpaca belongs.

Robertson, in his "America," avows the high estimation in which he held these early annalists, and enumerates the merits of such of them as he had occasion to quote; nor should it be forgotten that the author of the best epic poem in the Spanish language, bore a distinguished part in the conquest of both Peru and Chili. As eyewitnesses, these functionaries described the properties and habits of the Andes sheep in a plain and practical manner; and as the present essay is intended rather to show the benefits which, as agriculturists and manufacturers, we may derive from the adoption of the wool-bearing species, than to attempt any scientific arrangement or zoological classification, it may be necessary for the writer occasionally to avail himself of their testimony, in confirmation of such facts as he may seek to establish.

On no two points do the early writers on Peru agree so perfectly, as in reference to the number of the species into which the Andes sheep were divided, and the purposes to which the natives applied them. They state that there were four kinds, two tame and two wild; and this point is now too well ascertained to admit of a

---

* Zarate says, that after Pizarro had been deserted by the greater part of his companions, who returned to Panama, he was left on a small and uninhabited island, where his people were obliged to subsist on shell-fish and snakes. Exploring the Peruvian coast in their fragile bark, the persevering adventurers landed upon a convenient spot, where they procured from the natives several country sheep and other supplies. This was the first time these animals were seen, and their flesh eaten, by Europeans.

HISTORY OF THE ALPACA.

doubt. When the Spaniards reached the central declivities of the Cordillera mountains, they found them inhabited by various tribes of Indians, governed by Incas, or Emperors. The dominion of these princes was endeared to their subjects by the wisdom and benevolence which they displayed in the exercise of regal power, and more especially by their patronage of agriculture and the useful arts. In economical husbandry, and such other pursuits as conduce to the sustenance and comforts of man, as well as in the smelting of ores, and working of gold and silver, the Peruvians had, in fact, attained a degree of proficiency which astonished their invaders.

They were, however, chiefly devoted to a pastoral life, supplying the wants of their families from the produce of their sheep. Of these the Incas had large and numerous flocks as part of the regal establishment, the llamas of which were employed as beasts of burden; while the wool of the alpaca was distributed among the poorer classes living under the cold climate of the mountains, where the cotton plant does not grow. These flocks were divided according to their colour, and left to range on distinct grounds; great care being taken to class the young ones as soon as they were in a situation to leave their dams.* By this means, as far as possible, the flocks retained one uniform colour, each of which had its corresponding name, the white and black being most esteemed.

The growth of wool, it thus appears, was under the immediate patronage of the Incas, and it is only reasonable to suppose that they employed every possible expedient to improve the breeds of their sheep. The Priests of the Sun,† a rich and powerful body, also possessed numbers of these useful creatures, some of which were reserved for their own uses, and those of their dependents; while others were set apart for sacrifices to the presiding deity, and immolated at stated periods. In these acts of their religion, Acosta says that the priests paid particular attention to the colour of the victim, varying it according to the season and the object of the sacrifice.‡ The number of these animals kept by the ancient Peruvians must, consequently, have been great, in some degree proportionate to the numerous tribes whose wants they were destined to supply. Under the patriarchal government of the Incas, the poorer classes were chiefly dependent upon them for raiment and food; and they had, besides, to support and clothe armies of fighting men, labourers, and a host of menials. It was also part of their duty to provide against the contingencies of famine, by establishing suitable magazines of provisions. Of the number of the flocks kept by the Peruvians in ancient times, some idea may be formed, when it is taken into account that, after the enormous desolation caused by conquest had ceased, the aboriginal inhabitants were still estimated at from seven to eight millions; whereas they do not now amount to so many hundred thousands.

The comparatively small size of Peruvian sheep, as well as of the vegetable forms by which they are surrounded, clearly indicates that the climate of the Andes is not favourable either to animal or vegetable growth.

---
* *Inca Garcilasso de la Vega, Commentarios Reales del Peru*, lib. v. chap. iv.

† One of the earliest laws passed after the conquest, was, that all lands, cattle, and valuables, belonging to the Priests of the Sun, or in any other way devoted to sustain the worship of idols, should be confiscated to the crown, and heavy penalties were denounced against any person retaining or secreting the same. The most valuable flocks were apportioned to those superior officers who held command, and literally wasted, being at first often used to make tallow.

‡ *Natural y Moral de las Indias*, lib. iv. cap. 41.

It has also been remarked, that there the human species is subject to the same rule: man decreasing in bulk and stature in proportion as he dwells near the mountain summits. In Peru, the winter sets in towards June, and is severely felt on the highlands, where the snow remains upon the ground six, and in some places eight months in the year.

As soon as the narrow and green strip of land, bordering upon the Pacific, is passed, the traveller begins to ascend the slopes; and when he attains the first table-land, observes a complete change in the climate and the appearance of vegetation. Except in the *yungas*, or hollows, where an alluvial soil has been collected, and where the Indian plants his sugar-cane, *banana*, and esculent roots, the country wears a naked and barren aspect.

Here, at an elevation of from 8,000 to 12,000 feet above the level of the sea, the Peruvian tends his alpacas and llamas, allowing them to range at the foot of the snowy cliffs called *punas*, or to wander on the *paramos*, or heaths, where they derive subsistence from the moss and lichens growing on the rocks, or crop the strong grasses and tender shrubs which spring up upon the flats, favoured by moisture. On these commons the animals may be said to shift for themselves, exposed to all the rigour of the elements, and receiving no food from the hand of man. The shepherd only visits them occasionally; yet such are their gregarious habits, that the members of one flock seldom stray away and mix with another, being kept in a good state of discipline by the old ones, who know their own grounds, and become attached to the place of their nativity, to which they return at night, evincing an astonishing vigilance and sagacity in keeping the young ones together, and free from harm. Hence there is no need of their being marked; and so great is the intelligence of some *punteras*, or leaders of a flock, that a more than ordinary value is, on this account, attached to them by the owner, part of whose duties they perform.

The point nearest to the equator at which Andes sheep were originally noticed, is Rio-bamba, situated in latitude 1° 38′ S., about ninety miles S. W. of Quito, and not far from the snow-capped mountain of Chimborazo. The town stands 11,670 feet above the level of the sea, to which elevation the temperature of the air corresponds. In this tropical region, and consequently on a spot where excessive heats might be expected during the month of August, the two Ulloas remarked that, towards evening, the thermometer regularly fell two or three degrees below the freezing point, and next morning rose eight or twelve above it, which would indicate that, at a certain elevation, no land is exempt from the dominion of frost. On the contiguous mountain of Pichincha, where the Spanish commissioners performed part of their astronomical labours, they found the cold so intense, that in the little hut which served them for shelter, crowded as it was with inmates, and lamps constantly kept lighted, sometimes each person was obliged to have a chafing-dish, constantly supplied with burning charcoal, near him. The feet of the Europeans were nevertheless swoln, their hands covered with chilblains, and their lips so tumid and so much chapped, that the mere effort to speak often caused the blood to gush.*

Although, from the point above mentioned, across the equator, the climate becomes milder, and vegetation more abundant, it has been remarked that the wild species do not pass the line, but continue stationary there—a phenomenon for which some Peruvian writers have

---
* Voyages, book v. chap. ii. *Eng. Ed.*

HISTORY OF THE ALPACA.

endeavoured to account, by alleging that the *ichu* plant, a coarse grass, and the favourite food of both the tame and wild species, does not extend further towards the north.* This, however, is by no means a satisfactory reason; for it is not proved that this kind of herbage, peculiar to the central division of the Andes, reaches to the southern extremity, the direction in which the guanaco has migrated, even descending to the remotest plains in search of food. Naturally timid and solitary as this animal is, unobstructed by rivers, and even undeterred by precipices, it has nevertheless penetrated to the furthest parts of the Patagonian coast, as far as the low and swampy lands of *Tierra del Fuego*, where herds of them were noticed by our early navigators, and recently by Captains King and Fitzroy, of the Adventure and Beagle, whose crews found a seasonable and salutary refreshment in the guanaco meat furnished by the Indians. Along the low and broken coast, extending almost as far as Buenos Ayres, numbers of guanacos are observed pasturing, frequently on forest tracts, and ranging thence to the summits of the nearest hills, disappearing on the *pampas*, or inland plains, and again showing themselves near the frequented passes on the eastern declivities of the Andes.

It has been remarked by physiologists, that the size of animals is usually adapted to the nature of the country which they were born to inhabit. This is not the case in the present instance; and whether we consider the great extent of the Andes mountains, their stupenduous forms, the immense elevation of their summits, or the severity of the climate prevailing upon them, the more shall we be astonished at the diminutive size and delicate frame of the quadrupeds dwelling in those secluded recesses. The woolly natives, nevertheless, possess a hardiness of constitution, and a peculiarity of structure, admirably well adapted to the nature of their birthplace. There, during half the year, snow and hail fall incessantly, whilst in the higher regions, as before noticed, nearly every night the thermometer falls below the freezing point, and the peaks consequently, are constantly covered with an accumulation of ice. The wet season succeeds, when flashes of lightning traverse the clouds in rapid succession; the thunder rolls through the firmament in grumbling and prolonged peals, followed not by showers, but by torrents of rain, which, after collecting, fall headlong from the rocks, or pour into the crags and chasms, leaving the slopes bare of soil, and spreading desolation wherever they pass, till at length the stream is lost in some lake, or serve to swell the head waters of a river.

It is astonishing that the temperature of the air on mountains so peculiarly situated, and exposed to the full blaze of a vertical sun, should be so much chilled as almost to present the desolate aspect of the Arctic regions; and yet such are the tracts of land upon which the Andes sheep abound and thrive—the flocks, more especially those of alpacas, being still, comparatively speaking, considerable in the vicinity of Rio-bamba, where the inhabitants evince a great aptitude for woollen manufactures, and carry on a trade in the raw material. Of alpaca and vicuna wools the women knit stockings, with coloured cloaks, and also gloves equally ornamented.* *Ponchos*, or men's surtouts, are wove in colours, and of so delicate a texture as to be worth 700 dollars each. They are also used throughout Peru as a riding dress by the wealthiest ladies.

Pursuing their researches, the Spaniards ascertained that, at the period of their arrival, llama and alpaca flocks on the coast were kept as far as the fortieth degree of south latitude, and inland as far as the territory of the Araucanos, in which space they occupied the middle declivities of the Andes, facing the west, wherever population was concentrated. Alonso de Ovalle, a jesuit, and a native of Chile, in his *Historica Relaçion del Reyno de Chile*, (Rome, 1646,) says, that in the capital of Santiago, llamas formerly had been used to carry wheat, wine, and other articles, and also to bring water from the river to the houses. For many years this drudgery has been exclusively performed by mules and asses.

Along the extended range above named, the tame breeds were left to browse. The sheltered part of a hill, the bottom of a dale, or the furzy heath, were their favourite haunts. There they picked up their scanty and scattered food, under the lower boundary of the snow, ascending as it disappeared from the surface. Sometimes they fed on the mosses which fringe the rocks, and plants growing on the hillocks, or would descend the slopes and enter the *ichuales*; while, in the higher and more secluded regions, reaching nearly to the summits of the lofty chain, as well as on both sides of the double line which it assumes in Peru, dwelt the vicuna and guanaco in a wild state, and far from the abode of man, hunted only for their flesh and skins.

As regards the historical part of the subject, the preceeding outline may suffice. On the score of zoology, it may be proper to remark, that Cuvier places the genus *Camelus* at the head of the family of the hornless ruminants, and next in succession the llama, which term he applies as the generic name of that form of *Camelidæ* by which the New World is distinguished; whereas Illiger separates Andes sheep from the latter, ranging them in a distinct order, to which he gives the name of *Auchenia*, in allusion to the length of the neck. Whether this is a property sufficiently distinctive to warrant a separate division, this is not a place to enquire. The elder naturalists agree in their classifications of the llama race. Hernandez and Marcgrave call it *Ovis Peruana*, and Gesner, *Ovis Indica*; but, as will be hereafter seen, a modern naturalist, and one of the highest eminence too, gives it another and rather an extraordinary classification. Whatever may be the determinate structure, or the precise peculiarities, establishing the unity of *genus*, agreeably to the beautiful theory on the correlation of forms taught by Cuvier, his arrangement seems the most appropriate, as the llama unquestionably possesses some properties peculiar to the camel, at the same time that it is marked by others perfectly distinct.

From the remotest period to which the Peruvian records extend, it appears that the aborigines not only employed the llama and alpaca in their military and domestic service, as the Arabs do the camel, but also used them for food and clothing. The llama was principally destined to carry burdens, although, compared with the African drudge, the difference in size and strength is considerable. Its load never exceeded 150 lbs., with which it was not required to travel more than three leagues per day, whereas in the working part of the

---

* *Ichu (Jarara—class. Monandria Dygynia—Planta capitosa, bipedalis—foliis involutis subulatis—spicâ paniculata.)* Of the gramineous tribe, grows high and when at its full size resembles a reed, the stem strong and erect. It is found on the *punas*, or bleak and unfrequented spots of the greatest altitude; and when young serves to feed cattle, and full grown is used as fuel for cooking, in those barren regions where the dung of the llama and alpaca only can be had. In the Huancavilica mines it was formerly used in the furnaces, to separate the quicksilver from the ore. The Indians also use it for thatching. The llamas and alpacas are fond of the green tops; and when feeding in an *ichual* or *ichu* field, their bodies are hidden, while their heads are seen gracefully towering above their food.

* Such as Verona serge, twilled *ratinette*, Astrakan cloth, &c.

twenty-four hours, the camel journeys double that distance, with 800 lbs. or more upon his back.* For this difference the Peruvians made up in the greater number of their beasts of burden, one drove sometimes exceeding 500 head, whose subsistence on the road was entirely left to chance. Neither whip nor goad was used to urge them on. One llama, older and more experienced than the rest, led the way, the others following irregularly but quietly after. Owing to its docility and knowledge of its keeper, this animal evidently requires less training than the camel. It needs no rein—not even a pack-saddle—so long as the panniers are well poised.

Although it must be acknowledged that the llama and alpaca possess some properties in common with the camel, they are nevertheless separated from him by various and strong marks of distinction. Independent of size and figure, as regards anatomy the absence of the hunch, and in them the callosity on the sternum, constitute striking points of difference. The structure of the foot is dissimilar; their horned hoof, almost resembling the talons of a bird of prey, being very remarkable. In the dental arrangement there is also a variation: in the upper jaw of the Andes sheep there are no incisors; while the teeth, ranged underneath, are differently shaped, sharper, and stand in another position. On each side of the upper jaw they have a canine tooth more than the camel, and generally the osteological character of the head differs. In them the incisors project full half an inch from the muzzle bone, so as to meet the pad fitted above, by which means, and with the aid of the tongue and cleft lip, they are not only enabled to draw together and clip short grass upon the ground, but also with their long neck, pointed muzzle, and the oblique posture which the head can assume, to reach herbage growing on the ledges and in the interstices of rocks seven feet high, as well as the tops of hedges and tall shrubs. Their teeth are at the same time so strong, and interlock in such manner, that they easily crush and masticate vegetable substances too hard and tough for ordinary cattle.

In the formation of the stomach there is, however, a great similarity. The endurance of both hunger and thirst, so truly characteristic of Andes sheep, would indeed appear surprising if this peculiarity were not duly considered. In this respect they resemble the camel, which is not only capable of retaining water when once swallowed, but also of producing it—an important part of the animal economy, equally remarkable in his allied species of the New World, which have been noticed to abstain from drink so long as they are provided with succulent herbage. In them the stomach is divided into compartments, some of which are destined to hold food, while others serve as reservoirs for fluids; and one is struck with astonishment at the variety of the component parts of this double receptacle, as well as the complexity of their structure.

The cellular developement of that part of the stomach, which in the llama and alpaca is appropriated to the retention of fluids, together with other capabilities of the internal machinery, seem in fact to be more perfect than in the camel. Hence, besides being formed by nature on so economical a scale as to require the scantiest supply of food, and also the smallest possible quantity of drink, when ruminating they have the power of not only forcing the fluid back to the stomach, in order to assist digestion, but also to the mouth, in such a manner as to allay thirst. The apparatus by means of which the process of mastication is performed, appears equally admirable. Naturalists who have had the best opportunities of observation, admit that each variety has a peculiar power to extract and appropriate for its own use, and in the way of nourishment, the constituents of that food which it gathers; and hence the choice ought, as much as possible, to be left to itself. Any thing which it eats with repugnance, is sure to disagree.

Their patience under the uneasiness arising from the want of drink, exceeds that of the camel. The weary traveller over the Arabian desert is said to smell the well or pond at the distance of half a league; and, in the hope of refreshing his wasted strength, he evinces impatience to reach it. In South America the mule perceives water at a still greater distance, and quickens his pace to approach it; but if a llama, or an alpaca happens to be in company, the animal remains perfectly unmoved, although it cannot be supposed that the olfactory nerves of either are less powerful than those of the camel or mule.

Between Peru and Chili intervenes the desert of Atacama, which it takes several days to pass, and where in the dry season water is not to be had. The loaded llama nevertheless crosses without difficulty; and, if it had not been for his services in former times, no commercial intercourse by land could have been kept up between those two neighbouring countries. The Spaniards, in fact, could not have conquered Chili—at least at the period and in the manner in which they did—if they had not been assisted by llamas in carrying their luggage and water across the mountains.

In Peru and Chili the horned cattle and European sheep, which have multiplied to an enormous extent, while the llama and alpaca have declined, require the occasional mixture of salt with their food, it being ascertained that without it they do not fatten. So fond are they of this luxury, that they frequently roam to a great distance in search of the *salitrales*, or salt-licks, and when satisfied return home. Andes sheep do not evince the same eagerness for salt, and consequently never experience the inconvenience arising from the artificial thirst which it occasions. This abstinence from drink in them may, however, contribute to render their supply of milk scanty; of which, as it barely suffices for their own offspring, the Indian never deprives them.

Nothing better exemplifies the bounty of the Creator to man, than the gift of such serviceable animals as the camel and the llama. Had it not been for the former, the Arabs never could have been able to cross the parched and sandy deserts which intersect their country; nor, without the aid of the latter, would the Peruvians have been in a situation to ascend the steep and rugged sides of the Andes, with loads disproportioned to the strength of their delicate frames. The aborigines of Peru, however, differed materially from the Arabs both in habits and pursuits, and were besides gifted with two useful animals of the same race, instead of one. The

---

* The Bactrian, or two-hunch camel, carries from 12 to 1400 lbs. weight. In this race, perhaps more than in the other, we see a perfect exemplification of that curious and useful provision, ordained by nature, by means of which this animal is enabled to receive, in separate cells of the stomach, such surplus food and water as may not be immediately wanted to support life. This reserved nutriment, which the camel instinctively stores up, will last for some time, thus enabling him to travel over the barren hills where none can be had; another beautiful illustration of that adaptation of structure, so remarkable in the animal kingdom, to suit the conditions arising out of the physical peculiarities of particular regions. The anatomical and external bodily characteristics of the alpaca, show that it is equally well suited for its Alpine home. Andes sheep are marked by peculiarities adapted to the elevated region where they live and thrive, and where the air is extremely keen and rarified, and water scarce. Their exterior physical endowments indeed show that they were destined by Providence only for hilly grounds. In its habits of life the alpaca resembles no other animal, except its *analogus*, the llama.

Arabs have always been represented as a wandering people, accustomed to perform long journeys in order to exchange the productions of their soil for those of others. The children of the Incas, on the other hand, were stationary, and fond of home. Usually they led a pastoral life, and had carried agriculture to such a pitch, that on the broken ground occurring on the declivities of their mountains, with the alluvial soil brought from afar on the llama's back, they formed artificial terraces or *parterres*, having the appearance of hanging gardens. On these they sowed their *quinua* (millet) and maize, reared the *axi*, (red pepper,) or planted *ocas, arrucachas, yucas, camotes, papas*, and other edible roots.

The Quechuans and Aimuraes, the most numerous and civilized tribes inhabiting the slopes of the central Andes, were essentially shepherds and manufacturers. On the *plateaux* or level spots open to the western exposure, population was formerly so much concentrated, agriculture so far extended, and the flocks so numerous, that economy became necessary in the distribution of those lands, which are now little else than a trackless wilderness. The chief care of the natives was devoted to their sheep, which formed the standard of their wealth; and not only enabled them to feed and clothe their families, but also to carry their surplus produce to market. A primitive law still exists in Upper Peru, in virtue of which the llama and alpaca, at the valuation of three dollars each, are made a legal tender, and common sheep at four rials. This sensible enactment is derived from the Incas.

The Peruvian mountaineers treated their domestic animals with great kindness, as if sensible of their utility, and grateful for the predilection with which Providence had viewed them. Parsimonious in their habits, their sheep were always turned to the best account; and, on the arrival of the Spaniards, so striking was the contrast between the condition of the two tribes above named, and that of the Araucanos, Puelches, and Patagonians, who were hunters and fishermen, and as such continue to the present day, that it is only reasonable to attribute the higher degree of civilization which they had attained, and the comparative comforts enjoyed by them, to the possession of the llama and alpaca, coupled with the pastoral life which their owners led. With us, the sheep is a connecting link to man as regards support. Half the meat brought to our tables is obtained from that kind of stock, and so is the largest portion of our clothing. In exactly the same manner did the central Peruvians find provision in their flocks. The hunting tribes, dwelling towards the south, knew only the guanaco, on the flesh of which they fed, and of the skin made a warm quilt; but, to procure these requisites, they were obliged to be constantly roaming about on the wilds.

Frequently the llama, but preferably the alpaca, was a pet in the Indian's cabin; and certainly in intelligence both animals are superior to other ruminants. As regards patience and resignation they are equal to the ox; while in point of sensibility they surpass every other quadruped. The size and shape of the eye indicate a strong and quick sight, as well as a peculiar capability of bearing the reverberation of the sun's rays, in like manner as the camel resists the glare of the sands, which in man so often produces ophthalmia. Resembling the best breeds of our sheep, the llama and alpaca are small boned and well proportioned in their quarters; the back is broad and straight; the eye lively and prominent; the countenance open, and one of them, at least, is abundantly supplied with wool.

The organization of both is adapted not only to the nature of the ground which they were destined to tread, but also to the climate under which they were formed to live. The sole of the foot is guarded by a cushion, and the toes armed with hard and curved nails, by which means, and the great mobility of their limbs, they are enabled to climb a steep and craggy hill with agility and security. They may be justly called Alpine animals, and in point of activity will bear some comparison with the *Capridæ*, or goat tribe; while in their flesh, skin, and general appearance, they are not unlike the deer.

How the Peruvians originally became possessed of their domestic breeds, or whether the llama descends from the guanaco, and the alpaca from the vicuna—questions repeatedly started by naturalists in Europe—at this late period could only be a speculative enquiry; and in this respect the traditions of the natives afford no assistance. In their tame as well as in their wild state, the four species herd separately, and when left to themselves do not intermix. Although, from the several points of resemblance between the guanaco and the llama, it is possible to conceive that, by care and cultivation, accompanied by a change of climate and pasture, the former might improve in figure, as well as in the colour and quality of its coat, eventually becoming what the llama now is: there does not appear to be any probability that the vicuna was the parent stock of the alpaca. The difference in size and structural formation, the dissimilarity in the wool, and other points of disagreement, present insuperable difficulties; and it is not to be forgotten, that in neither instance has any change in climate or pasture occurred.

Nor can it be argued that a new race has been obtained by means of crossing. The llama and alpaca certainly can be induced to intermix, and of this union there are frequent examples in Peru. From the alliance a beautiful hybrid, in fact, results; if possible, finer to the eye than either parent, and also more easily trained to work, but, like the mule, it does not procreate. Numbers of this intermediate species may be seen in Upper Peru, where they are called *machurgas*, a corruption of *machorra*, a Spanish word meaning a barren sheep, and equally applied to woman:—

" Y sus machorras obejas,
Vengan á sus parideras,
Con que doblen su ganacia."
CERVANTES, *Galatea*, lib. iii.

Wishing to confirm my own opinion upon this curious and important point of natural history, formed from intelligent communications more recently received from Lima, I wrote to General O'Brien, an observant Irish gentleman who resided twenty years in Peru, ten of which he served as aide-de-camp to San Martin, the Liberator, and was actively employed during the War of Independence—a great traveller on the Andes, and besides a landed proprietor and miner in the district of Puno. Subjoined is his reply, dated Liverpool, June 6, 1841:—

"You ask me whether the alpaca is still used in Peru as a beast of burden. I answer that it is, but not generally, and only by the poorer class of Indians, who do not own many llamas. There is, however, a beautiful animal produced between the llama and alpaca, much handsomer in form and figure than either, and also better adapted for work, but it does not breed. A few days ago, at the Earl of Derby's, I saw a young one had between the llama and alpaca, and I have also seen another fine animal of the same kind at Mr. Edwards' of Halifax. In Peru we call them *machurgas*, and these are the animals I principally used at my mines to bring down the ores from the mountains."

From the sterility of the hybridous race, it would follow that the alpaca is a distinct variety of the llama tribe, differing as much from its allied species as the horse does from the ass; and, consequently, that the two domestic animals of the Peruvians were not brought to their present state by means of crossing. Their intermixture is a modern expedient, adopted only since the time of the Spaniards. It is a rule of the vital economy, that life springs only from life, and every being is consequently endowed with the property of generating an offspring, inheriting a nature similar to its own. Where the species vary, this rule ceases to act; whence, although possessing a strong physiological resemblance in many important points of their organization, there must necessarily be some material difference between the llama and alpaca in the functions of generation, which it is more than presumable equally extends to the wild species, and that difference produces an irregularity at variance with the law of nature, constituting an essential condition of life.

The state of perfection to which the Incas had brought the two domestic breeds of Andes' sheep, naturally suggests the query, why, in the same provident spirit which guided their actions, they did not undertake the domestication of the vicuna, whose wool was always held in higher estimation than even that of the alpaca. The answer is obvious. The Peruvian emperors had no occasion to domesticate and bear the charge of breeding vicunas, even had it been practicable. By means of their periodical hunts, of which the early writers have left us ample descriptions, they had an economical mode of obtaining from the animal all they wanted for their own clothing, and that of the privileged orders, besides providing a healthy and favourite amusement to their subjects, during which feats of dexterity were, in all probability, noticed and rewarded.

Neither is any mention made of the Priests of the Sun sacrificing the vicuna. In their estimation, the alpaca held the same place as the white elephant among the Siamese, and white cow with the Hindoos. Hence it was considered an animal of good omen; and as it formed part of the mystic doctrines of those priests, that the Deity could only be propitiated by the spilling of blood, the value and efficacy of the offering, in his eyes, was supposed to be enhanced by the higher estimation in which the victim immolated was held on earth. Acosta affirms, that in the capital of Cusco an Andes sheep was sacrificed every morning, probably at sunrise, as a greeting on the great luminary of day, to whom their adorations were addressed as being the only universal source of light—the being who creates and sustains all things. During some of their solemn festivals, we are told that the Peruvians immolated as many as one hundred of these animals at a time, an additional proof of the magnitude of their flocks.*

CHAPTER II.

ALPACA WOOL AND MEAT.

Of the two domestic species, obviously the alpaca is the preferable one for our adoption. Whatever may be

* For ordinary sacrifices, black sheep were more frequently used. The lamb was also a favourite offering; but, from an economical principle observable in the laws of the Incas, the female was never sacrificed until past bearing.

the classifications of naturalists, the alpaca is essentially a wool-bearing animal; and of the fineness and softness of the textures made from its fleece by the ancient Peruvians, we find ample testimony in the works of several of the early annalists, some of whom acknowledge that, in delicacy of weft, they exceeded any cloths at the time manufactured in Spain. The same authorities speak in the highest terms of the beauty and permanency of the vegetable dyes used by the Peruvians, as well as of the artistic skill of the designs, and the trueness with which their cloths were made. Spinning and weaving, indeed, formed part of the domestic employment of both men and women; and, as an encouragement, the Incas kept public establishments, in which the art of fine weaving was taught. D'Orbigny affirms that textures of both wool and cotton, extremely fine, and wove with perfect regularity, are still found in the *huacas* or sepulchral monuments of the ancient Peruvians;* a proof not only of their advancement in the art of weaving, but also of the durability of the materials used by them.

The capability of alpaca wool being converted into articles of fine texture, is now established by the experience of our own artizans, as well as those of France. Fancy goods made from this material, and having a superficial lustre resembling that of silk, for several years past have been selling in the London and Paris shops, and are very generally introduced into Germany and other countries. In quality this wool differs from that of ordinary sheep, exceeding it in length, softness, and pliability. The staple of English wools is seldom more than six inches long, whereas that of the alpaca averages from eight to twelve, and sometimes reaches twenty, acquiring strength without being accompanied by coarseness the reverse of which occurs in our woolly tribes. Each filament appears straight, well-formed, and free from crispness, and the quality is besides more uniform throughout the fleece. There is also a transparency, a glittering brightness upon the surface, which gives it the glossiness of silk, considerably enhanced when it comes out of the dye-vat. It is distinguished by softness, essential properties in the manufacture of fine stuffs; and being exempt from spiral, curly, and shaggy portions, when not too long it spins easily, and yields an even and true thread. Neither is it liable to *cotting*, which renders wool adhesive, and causes it to form knots, difficult

* *L'Homme Américain de l'Amerique Meridionale, considéré sous ses Rapports Physiologiques et Moraux.* M. Alcide D'Orbigny, a distinguished naturalist, was commissioned by the Paris Museum to explore the interior of Patagonia, Chili, and Peru, where he remained from 1826 to 1833. The principal part of this time was spent in Upper Peru, (Bolivia,) that immense *plateau* studded with lakes, and presenting a surface nearly equal to that of France. This elevated region contains the greater part of the population, purely Indian, now remaining in the Peruvian division of the Andes, and at the same time presents scenery of a novel and striking character. Here M. D'Orbigny resided several years, joined the natives in their struggle for independence, and became intimate with the various tribes of Aborigines, whose confidence he gained, and whose character and habits he had the best opportunity of studying. Hence the result of his observations on their physiological and moral qualities, forms one of the most interesting pictures that possibly can be contemplated. His researches in natural history are also most extensive. His voluminous work in large quarto, filled with beautiful illustrations, has for the last ten years been in the course of publication at Paris, and has now reached the seventy-third *livraison*; but the learned author has not yet entered on the zoological part, wherein he proposes to describe *Les Animaux Ongulés*, among which he classes the Andes sheep. *Les Nouvelles Annales du Museum*, (Paris, 1835, Tom. 3,) contain an elaborate report on his promised works, drawn up, after a careful inspection of the author's MSS. and specimens, by Messrs. Geoffroy Saint Hilaire and Blainville, two eminent French naturalists—a labour which they performed by orders and under the auspices of the French Institute.

ALPACA WOOL AND MEAT.

to unravel in the combing process. It is not injured by keeping, nor does it lose in weight.*

Avowedly the climate and pastures of Peru are far from being favourable to the quick growth of wool; and hence, by the natives, the shearing season is not regularly observed. In some places the alpaca is shorn annually, and in others every second year; but frequently the young ones run three summers with the same coat upon their backs. In performing the operation there is also great negligence, and a want of care in sorting the colours and qualities, which are indiscriminately huddled into one bag. The fleeces range from 10 to 12 lbs. each, whereas those of our full-sized sheep seldom go beyond 8 lbs., and the small breeds 4 lbs.† From the larger size of the animal, and the increased surface consequently covered, the alpaca necessarily yields most wool; and it had already been ascertained that on our soil the weight improves. At the Royal English Agricultural Show, held at Liverpool in July 1844, a sample of black wool was exhibited, taken from an alpaca belonging to the Earl of Derby's flock, the staple of which appeared to be about a foot long; when his lordship's farm-agent expressed his conviction that the same animal had then 17 lbs. upon its back.

This is a greater result than was ever obtained in Peru. An improvement in both weight and quality might indeed naturally be expected, it being an undeniable fact that our country is much better suited to the growth of wool than the barren and inhospitable *lomadas*, or slopes of the Peruvian Andes. With us more care would also be taken of the stranger, and in cases of disease he would experience skilful treatment. Greater attention would, at the same time, be paid to the sorting and packing of the wool, which is best when the staple does not exceed from six to eight inches, the process of carding and spinning being more tedious when it goes beyond that length. By regularly shearing the animal every year, we should also obtain wool of one uniform staple; an advantage which we can never expect, so long as we are obliged to import it.

As, however, the breeder looks forward to returns, not only on the fleece, but also on the carcass of his sheep, it may be proper to say something respecting the quality of alpaca meat, which, owing to the extravagant prices at which these animals have been bought in England, and their being hitherto considered as objects of curiosity rather than speculation, has not, I believe, been tried with a view to ascertain its flavour and properties. That the ancient Peruvians used alpaca and llama meat much in the same manner as we do mutton, is beyond all doubt established. Acosta affirms that, in his time, it was considered good, and when young he himself thought it as delicate as any food that could be eaten; further remarking, that of the flesh of both animals the Indians made jerked meat, which kept well and was much liked.‡

Inca Garcilasso assures us that, when he wrote, llama meat was considered better than any other in use—tender, wholesome, and savoury; and that the flesh of the young ones, four or five months old, was by physicians recommended to sick persons in preference to fowls. The alpaca, he proceeds to say, is not adapted to carry burdens, but usually kept on account of its flesh, which is nearly as good as that of the llama, and its excellent and long wool.* Zarate declares that llama and alpaca meat is extremely wholesome, and as palatable as that of fat sheep in Castile. He further remarks that, in his time, there were shambles in the Peruvian towns where it was sold, although this practice had only been adopted subsequent to the arrival of the Spaniards; as previous to that period the natives were in the habit of killing a sheep, each in his turn, and sharing the meat with his neighbours.†

The quality of alpaca meat could not indeed fail to be good, when the cleanliness of the animal, the nature of its food, and the neat and delicate manner in which it feeds, are considered. Andes sheep eat nothing but the purest vegetable substances, which they cull with the greatest care, and in habitual cleanliness surpass every other quadruped. With their flesh the Peruvians, to this day, prepare a jerked meat, called *charque*, which, stewed with rice, or onions and tomates, makes an excellent dish. On their farms, it in fact holds the place that bacon does on ours, and also serves for a sea voyage. To prepare it, the meat is separated from the bones and cut into long slips, with a due proportion of fat adhering to each, and all the coarse bits rejected. In this state it is slightly salted, dried in the sun, and then smoked; by which process, however, it becomes so hard and dry, that it requires steeping in water for several hours before it is used. Andes sheep eat very much like the venison ones cured in North America, and certainly the dried tongues are superior to those of the reindeer.

In Lima, Ulloa assures us that a haunch of llama and alpaca is much esteemed, and, at a particular season, sent down from the highlands, prepared in such a manner as to keep without the aid of salt. For this purpose the meat is exposed forty-eight hours to an intense frost, when it becomes thoroughly congealed. In this state it is preserved sweet for a month, or more, if not exposed to a damp atmosphere, or the immediate action of the sun; and, when cooked, is found to have lost little of its freshness and flavour.‡ When the alpaca is of a proper age, and well fed, the cleft of the haunch is smooth and close, the meat small grained and rather mottled, the fat white and firm, and, from three to four years old, of full flavour. It is not a greasy but rather a juicy meat, and easily digested. The flesh of a full grown one is more nutritious than that of the yearling, although the latter is delicate and savoury, and would make an excellent ingredient for a pie.

In point of flavour, alpaca meat has, by good judges, been compared to North American venison, and even to our heath-fed mutton.§ The nature of pasture unques-

---

* A small bundle of alpaca wool, with a few locks of Australian mixed with it, was accidently thrown into a closet and forgotten. At the end of twelve months it was opened, when it appeared that the moths had nearly eaten up the Australian without injuring the alpaca wool.
† Sheep of the Shetland breed only weigh from 24 to 26 lbs. of mutton, and the fleece from 1 lb. to 1 1-2 lb. They are not shorn, the wool being pulled from the animal's back; and, when the shaggy and coarse parts are rejected, the quantity in each fleece, suited for knitting and weaving, seldom exceeds half a pound clear, and is worth from 6d. to 1s. 3d. per lb.
‡ *Historia Natural y Moral de las Indias*, lib. iv. cap. 41.

* Ibid. lib. viii. cap. 16.
† *Historia del Descubrimiento y Conquista del Peru*, lib. iii. cap. 2.
‡ Ulloa further observes, that a frozen calf is frequently sent down from the mountains as a present, and with care may be kept a reasonable time without any symtoms of putrefaction. By means of congelation the Peruvians also prepare their favourite food, called *chuno*. This is done by putting peeled and sliced potatoes into a sack, which is immersed in water, and the vegetable there left for three days, when it is taken out, spread upon a dry surface, and exposed to the action of frost. It is then put away in the store-room, and, stewed with meat, forms a kind of permanent dish.
§ As regards the flesh of the larger of the two wild species, I have it in my power to offer the following testimony from Mr. Darwin, the talented naturalist who accompanied the late surveying expedition in the Beagle, round Cape Horn :—"I have much pleasure in answering, as far as lies in my power, your enquiries regarding the guanaco. The first I killed was at Port Desire, on the coast of Patagonia; it weighed, without blood, en-

tionably affects both the quality and taste of meat. There is therefore every reason to expect that, fed upon our downs or heaths, the alpaca would yield a good and marketable flesh, thus increasing our supply of one of the necessary commodities of life. In this kind of stock the breeder would find another desideratum, and this is, the largest quantity of flesh on the least possible weight of bone. The quarters weigh from 35 to 45 lbs. In Peru this meat is not sold by the lb., but by the lump, and the price regulated by custom.*

In Upper Peru the aboriginal tribes continue to eat alpaca meat, and many Europeans on the spot prefer it to mutton or beef. From motives of economy the Indians, however, seldom kill any other than old stock for their own eating; but at their merry-makings a haunch of young llama, or alpaca, holds a conspicuous place on the table. Owing to the nitrous pastures and aromatic plants growing on the middle declivities of the Andes, which afford food to the European sheep, mutton there is good, partaking something of a game flavour, and the same causes may give to llama and alpaca meat a similar taste. To please the early missionaries, the Peruvians adopted the sheep brought from Spain, which now serve as a kind of appendage to their own flocks. The offspring of merinos, under an altered, and, as regards form, improved appearance, may be seen in many places browsing in company with llamas and alpacas, although not considered equal to more than one-sixth of their value. The Indians and Mestizos nevertheless prefer their old fare to mutton, in consequence of which, although plentiful, the use of the latter is almost confined to the natives of European origin.

As the skins of Andes sheep may be turned to some account, they ought not to be entirely overlooked. The first use made of them by the Spaniards was peculiarly national; for they prepared them exactly in the same manner as they still do the goat-skin at home, for the purpose of carrying wine. Zarate enumerates the difficulties and privations which Diego de Almagro experienced in conveying his people from Cusco across the mountains to Chili, principally owing to the want of water, to obviate which llama skins were filled with supplies, and carried on the backs of living animals of the same class. An alpaca skin is not so thick as that of the deer, but strong and pliant; and in an ingenious and manufacturing country like ours, might be appropriated to various useful purposes, among which bookbinding is not the least important.† The alpaca mantle has ceased to be worn in Peru; but the soft and warm pelt still affords a comfortable bed to the mountain tribes. The skins of the wild species are also found serviceable. The Patagonians and other Indians, living in an independent, and almost savage, state on the southern extremity of the continent, prepare the guanaco skin in such a manner as to shield them from the inclemency of the weather, turning the furry side as the season changes.

CHAPTER III.

APPLICABILITY OF THE ALPACA TO OUR SOIL AND CIRCUMSTANCES.

FROM the experiments already made, not only in the British isles, but also in several parts of Europe, we are now sufficiently well acquainted with the properties of the tame species of Andes sheep, to feel assured that they are hardy animals, and easily fed. From unquestionable authority, we also know that they were found in the highest degree useful by a race of secluded mountaineers, engaged in the peaceful occupations of pastoral and agricultural life, and who without them scarcely could have existed. Of the two kinds, the alpaca, as before stated, is evidently the most valuable; as, besides furnishing a wholesome and nutritious food, it yields a fine and glossy wool, which might easily be made the staple commodity of a new manufacture, and by thus opening another source of trade, help to remove that pressure which bears so heavily upon various classes in the community.

By trials commenced more than twenty-five years ago, it is equally placed beyond doubt that this animal may, without any great difficulty, be naturalized among us, and made to propagate; and every day the facilities and the efficacy of the scheme to adopt it, become more apparent. The hardy nature and contented disposition of the alpaca, cause it to adapt itself to almost any soil or situation, provided the heat is not oppressive, and the air pure. The best proof of its hardiness is its power to endure cold, damp, hunger, and thirst, vicissitudes to which it is constantly exposed on its native mountains; while its gentle and docile qualities are evinced in its general habits of affection towards its keeper.

No animal in the creation is less affected by the changes of climate and food, nor is there any one to be found more easily domiciliated than this. It fares well while feeding below the snowy mantle which envelopes the summits, and for several months in the year clothes the sides of the Andes. As before shown, it ascends the rugged and rarely trodden mountain path with perfect safety, sometimes climbing the slippery crag in search of food, and at others instinctively seeking it on the heath, or in rocky dells shattered by the wintry storm; at the same time that, when descending, it habituates itself to the wet and dreary ranges on the lowlands, so long as it is not exposed to the intense rays of the sun.

This peculiar facility of accommodating themselves to different climates and situations, so remarkable in the tame varieties, we also know distinguishes the guanaco, which, as I have already had occasion to observe, has in the course of time spread to the southern limits. In a communication addressed to me by Mr. Darwin, whose authority has previously been quoted, are the following remarks upon this subject:—

"Perhaps there is no animal in the world which, in its wild state, flourishes under stations of such different, and indeed directly opposite characters, as the guanaco. I saw them on the hot deserts near Northern Chili,

---

trails, or lungs, 170 lbs. Another, shot a few days afterwards, was estimated at a greater weight. These, and during the succeeding year many others, were served out on board H. M. ship Beagle as fresh meat, and were generally liked. The meat, as far as I can remember, was fine grained, not very dark, (perhaps of about the same colour as mutton,) rather dry, but not with the least bad taste or smell. I do not, however, think it would be considered of a very fine flavour; but, on the other hand, it must be remembered that the meat was tried in no other way (as I believe) except being baked in a ten gun brig's stove, and that it was eaten very fresh. Moreover, these animals, shot in this wild state on the desert plains, were not fat. I cannot doubt that the guanaco, if domesticated and fattened, would yield a meat which, when well cooked, would be decidedly good, although possibly not equal to beef and mutton."

* At first the Spaniards made soap and candles of the fat of Andes sheep; but the immense multiplication of horned cattle, has since furnished them with a readier and more abundant material.

† The greater part of the alpaca skins brought to this country are in the wool. The ancient Peruvians made sandals of them, which they always took off when fording a stream.

APPLICABILITY OF THE ALPACA TO OUR SOIL AND CIRCUMSTANCES.

where the climate is excessively dry; on the borders of perpetual snow, at the height of 12,000 feet; and on the rocky and bare mountains of the same country. They swarm in great herds on the most sterile plains of gravel, composing Patagonia. Formerly they were numerous on the grassy savannahs stretching on the banks of La Plata, where during half the year the summer is hot, and in the winter abundant rain falls: and lastly, the guanaco lives on the peat-covered mountains, and in the thick entangled forests of *Tierra del Fuego*, of which country the climate is far more humid and boisterous, and the summer less warm, than in any part of Great Britain. I could perceive no difference in the guanacos of these several regions. If the alpaca be the same species, or has the same constitution, as the guanaco, these facts regarding the range of the latter are interesting, as they show under what various conditions we might expect the alpaca to thrive. I will only add, that the guanaco so easily becomes tame, that young ones, caught and brought up at farm-houses, seldom leave them, although ranging at full liberty near their native plains."

During the reign of Philip II., at which period parcels of alpaca and vicuna wools were occasionally brought from Peru, and much admired, a project was formed to bring over these animals alive to Spain, with a view to their being naturalized, which was only defeated by the maritime war in which the Spaniards at that time became involved. The same design was revived by the French in Buffon's time;* but the attempt was not made till the days of Napoleon, when the Empress Josephine, under the advice of men of science, and through the instrumentality of the Spanish minister, Godoy, caused a mixed flock of thirty-six llamas, alpacas, guanacos, and vicunas, to be collected at Buenos Ayres, for the purpose of being shipped to France, as soon as a safe opportunity presented itself.

In 1808, at the period when the Spanish revolution broke out, the survivors of this flock, eleven in number, arrived at Cadiz, and were afterwards deposited in a menagerie at San Lucar, in Lower Andalusia. Here they were in a healthy condition, and had procreated, when the French army under Soult took possession of that part of the country. They were visited by the naturalists accompanying his division of the invading force, who reported that in the depot they found a female llama pregnant by an alpaca, and also three alpaca-vicunas, whose fleeces were finer and more abundant than those of the pure breeds.†

That in Peru alliances are frequently formed between the two domestic species, has already been stated; and, owing to the peculiar character of these animals, it is possible that a cross between them and the wild breeds might equally be obtained, as regards the one even with advantage to the fleece. But of what use would this intermediate race be, if, for reasons already explained, it is condemned to become a barren stock? The Spaniards were proud of their acquisition, thinking that they had thereby obtained a new race of wool-bearing animals, calculated to people their hills, and repair the loss sustained through the decline in their merino flocks. By the experiment of crossing, they, however, defeated the very object which they had in view, as the animals gradually died off without leaving any offspring; and after a lapse of thirty-five years, and ample proofs that the climate is by no means unfavourable, it does not appear that there are in Spain, at the present period, beyond half a dozen scattered Andes sheep, kept as curiosities.

These animals have lived to their full period on the lowlands of Spain, and yet found a congenial climate in France. They have bred at Hamburg,* and in England there have been repeated instances of births; but, unfortunately, several breeders fell into the error of crossing, by which means the propagation of the preferable race has, to a certain extent, been retarded. The few individuals, however, among us who commenced their experiments with pairs, or in very small numbers, have found the alpaca, whether imported or born on our soil, healthy, contented, and disposed to thrive; but the number hitherto has been too limited, and too much subdivided, to exemplify the benefits derivable from its more general adoption, or to point out the safer and more economical means by which so desirable an object may be attained.

The various possessors do not communicate with each other on the subject; neither have they made known the result of their observations to the public. It is, nevertheless, satisfactorily established that, far from manifesting any symptoms of deterioration, the wool of the alpaca improves on British pasture. The common sheep were originally natives of a warmer region than those where they now abound, and where the choicest breeds prevail. Why, then, shall not the same success attend the alpaca, when the difference of both climate and pasturage is so much in favour of the experiment? If that interesting animal can endure the bleak, boisterous, and chilling climate of its native mountains, why shall it not thrive on ours, where the elevation is less by two-thirds, and the winter comparatively mild?

Many of our northern hills would try the constitution of any sheep, and yet there the weather is never so inclement or so variable as on the Cordilleras of Peru. With so many advantages, why then shall not the alpaca have an opportunity of competing with the black-faced sheep, the only breed that can exist in those wild and inhospitable lands? Of the two, the stranger would fare best on scanty and scattered food, at the same time affording to the owner a far better remuneration. When ordinary sheep are removed from a cold to a warm climate, the wool becomes thin and coarse, until at length it degenerates into hair. This is the case with those taken from England to the West India Islands; whereas the merinos conveyed from Spain to Peru, and bred upon the Andes slopes, yield a fleece which, when well dressed, is preferred by the manufacturer to that of the parent stock.

As regards the alpaca, we bring a lanigerous animal from a dreary and barren situation to one equally well suited to its habits, and at the same time infinitely healthier and better adapted for feeding. The result, therefore, could not fail to be favourable. The atmospheric changes in our climate can have little or no influence on an animal constitutionally hardy and so well coated; and by the adoption of this stock we not only secure to ourselves a new raw material for our manufactures, but also an additional provision of butcher's meat.

If the animals take to the soil, and this, as before observed, they have done even in situations by no means well chosen, an increased weight of both fleece and car-

---

* In the supplement to his work on Natural History, may be seen a paper on the domestication of the vicuna in France, by the Abbé Bellard, in which the writer advises the ministry to employ a commercial house at Cadiz, in order to obtain the animals from Buenos Ayres, whither (he observed) they might be brought down across the plains of *Santa Cruz de la Sierra*.

† *Dictionnaire Classique d'Histoire Naturelle*, ad verb. *Chameau*.

* The late Mr. Ducrow's white alpaca, saved from the flames when the amphitheatre was burnt, had two young ones there.

cass must follow. An improvement in the quality of the wool may be equally looked for; it being abundantly proved that pasture has a greater influence on its fineness than climate. The staple, also, cannot fail to grow longer, if the animal has a regular supply of suitable food; and, for reasons already explained, this is more readily met with on our mountains than on those of Peru, where the flocks are exposed to great privations.

In other respects, the alpaca would prove an economical stock. It is freer from constitutional diseases than ordinary sheep, and less subject to those arising from repletion and exposure to rain; neither are its young liable to those accidents which befall the lamb. The mothers are provident and careful nurses; nor do the young ones require any aid to enable them to suck. Except at the rutting season, these animals stand in need of no extra attention; neither are they predisposed to take cold. In this respect, the alpaca is pre-eminently favoured by nature. Its skin is thick and hard, and, being covered with an impervious coat, it is not injured by moisture. Snows and storms never affect these animals. Unhurt they pass through the utmost rigour of the elements, and hence the precautions adopted by our shepherds on some bleak localities, with them would be superfluous.

Another remarkable feature in the alpaca is, that it does not often transpire; for which reason, and its peculiarly cleanly habits, the fleece does not require washing before it is taken from the back. Although often confined to regions, where

"Snow, piled on snow, each mass appears
The gathered winter of a thousand years,"

the alpaca is not subject to catarrhs, or to those disorders which disable the limbs. The chest being guarded by a callosity, or cushion, which comes in contact with the ground while the animal reposes, the vital parts are not injured should the flock be obliged to pass the night in a damp or unsheltered situation. Besides being free from the diseases incidental to common sheep, the alpaca is less exposed to what is called "outward accidents." The facility with which this animal escapes from the fatal consequences of a snow-storm is a valuable property. One shudders at reading the graphic description, given by the Etterick Shepherd, of those sudden and awful calamities which have so often overtaken the farmer in the Scotch Highlands, when

"The feathery clouds, condensed and furl'd,
In columns swept the quaking glen;
Destruction down the vale was hurl'd
O'er bleating flocks and wondering men."

I know not whether, in our hemisphere, the winters have become more severe than in ancient times; but since the well-known "Thirteen Days' Drift," supposed to have taken place in the year 1660, at which period so large a portion of the Scotch flocks was destroyed, and so many persons perished, it is a fact that we have had no less than thirty-six inclement seasons, during which the losses among sheep were incalculable. Nor have these misfortunes been confined to Scotland. The fall of snow which occurred towards the close of Febuary 1807, was so heavy in England, that in exposed situations the herds and flocks extensively suffered. Of the large number of sheep, on that occasion, overwhelmed in the Borough Fen, only 600 could be dug out alive, the rest being completely buried in the snow. Upwards of 2000 perished on Romney Marsh, and the desolation equally spread to other places.

In our islands, sheep are sometimes smothered by the snow falling down upon them from the hills, or perish in an accumulation of drift. Frequently they have not the courage, or the strength to extricate themselves; but from his greater size, boldness, and activity, the alpaca is better able to contend with the storm. In their own country, these animals have an unerring foresight of approaching danger, and, collecting their young around them, seek the best shelter which the locality affords. After a tempest seldom is one missing, although they are, as it were, left to themselves, and the country bare of trees. Nothing can be more interesting than to see a flock of Andes sheep overtaken by a storm, and crossing a valley, with the drift reaching to their very backs. Raising their heads in a bold and majestic manner, the old males take the first line, and by pushing through the barrier, or jumping upon it when resistance is too great, succeed in opening or beating down the snow, so as to form a path for the weaker ones to follow.

CHAPTER IV.

BENEFITS WHICH WOULD ACCRUE TO THE BRITISH FARMER AND MANUFACTURER FROM ITS NATURALIZATION.

In whatever point of view we contemplate the properties and habits of this animal, it will be found a suitable and saving stock on our hill farms. Flocks of them would not require the ordinary pasture lands to be set apart for their use; waste and unprofitable ones would suffice. With them gorse is a favourite food, and, when young, they devour it eagerly without being annoyed by the prickles. They would browse on the heath and wild grasses of our moors, consuming that herbage which sheep and cattle usually reject.* The peat-bogs, covered with Alpine plants, and now only half depastured, during one part of the year would afford sufficient sustenance to them. They would, consequently, in the least possible degree interfere with the live stock already located on our soil, and by no means diminish the supply of food reserved for it.

In former times the grazier relied more on the fleece than on the carcass of his sheep, for the payment of his rent and expenses; but German supplies have now so far superseded the use of our own wools in the manufacture of the better qualities of cloth, that its value has almost fallen below a remunerating price to the grower, and other fields of competition have, besides, been opened in Australia, Africa, the East Indies, and South America.† To the British farmer, who has wild and woodless tracts to stock, and whose object is to grow wool, the alpaca might, therefore, be made invaluable; as, besides a saving in keep and in care, the fleece is much heavier than that of common sheep, and the price nearly double. Its growth would also be less liable to competition, at the same time that it could not diminish the value of our native flocks. As before stated, alpaca wool resembles

* Florin grass, abundant on our hills of the greatest elevation, would furnish good food; and as it can, at all seasons of the year, be made into hay of a superior quality, abounding in saccharine, and more grateful to cattle than any other kind of grass, it would be an excellent winter fodder, should the snow remain upon the ground for any unusual length of time.

† The quantities of sheep and lambs' wool imported, were—From New South Wales, in 1842, 8,725,973 lbs., and in 1843, 11,942,605 lbs.—Van Diemen's Land, in 1842, 3,491,685 lbs., and in 1843, 3,993,040 lbs.—South Australia, in 1842, 690,396 lbs., and in 1843, 1,387,514 lbs.—Cape of Good Hope, in 1842, 1,265,768 lbs., and in 1843, 1,728,453 lbs.—East India Company's territories, in 1842, 4,246,083 lbs., and in 1843, 1,926,129 lbs.—Rio de la Plata, in 1842, 1,460,105 lbs., and in 1843, 1,879,653 lbs.

BENEFITS OF NATURALIZATION.

silk in appearance, as well as in many of the purposes to which it is applied, and with the consumption of silk only could it interfere.*

Formerly, plain worsted stuffs constituted the principal manufacture of Bradford. The ingenuity of the artist being, however, always on the stretch, during the late depression in the trade of that enterprising and industrious town, figured goods were commenced, and became so successful, that some makers were unable to produce a sufficiency to meet the demand. As a variety, the use of alpaca wool was attempted; but owing to the difficulty of spinning it, chiefly occasioned by the great length of the staple, some time elapsed before an adequate process could be discovered. The first person in this country who introduced a marketable fabric made from this material, was Mr. Benjamin Outram, a scientific manufacturer of Greetland, near Halifax, who, twelve or fourteen years ago, sold it at a very high price, in the form of ladies' carriage shawls and cloakings, as curiosities.†

No quantity of the wool existing in England, he was obliged to procure a small supply from Peru, and gradually the articles manufactured with it came into notice. In 1832, Messrs. Hegan, Hall, and Co., spirited merchants in Liverpool, convinced from their superiority that these new manufactures would erelong come into fashion, directed their agents in Peru to purchase and ship over to them all the parcels of alpaca wool they could meet with, and thus was laid the foundation of that valuable and growing trade in this article which has since risen up.

In 1834, the quantity of alpaca wool, reported as shipped at Islay, chiefly to England, was 5700 lbs., valued at 16 dollars per quintal, (100 lbs.:) in 1835, 184,400 lbs., at 18 dollars; in 1836, 199,000 lbs., at 23 dollars; in 1837, 385,800 lbs., at 20 dollars; in 1838, 459,300 lbs., at 25 dollars; in 1839, in the port of Islay, 855,500 lbs., and at Arica 470,000 lbs.—total, 1,325,500 lbs., at 30 dollars; in 1840, at Islay, 1,300,000 lbs., and Arica, 350,000 lbs.—total, 1,650,000 lbs., at 25 dollars; 1841—total, 1,500,000 lbs.; and in 1842, to the 9th of July, 1,200,000 lbs.—grand total, 6,909,700.

Since the tariff law came into operation our own custom-house returns are, from the 9th of July 1842 to the 5th of January 1843, 2,43299 lbs., and in the year ended January 1844, 1,458,032 lbs. Agreeably to the preceding estimates, from the year 1834, when this trade commenced, up to the commencement of 1844, we must have received 8,657,164 lbs., chiefly imported into Liverpool; but intelligent merchants on the spot, and engaged in the trade, pronounce the Peruvian returns to be defective, confidently asserting that, according to data in their own possession, the quantities arrived within the period above specified was much greater than that set down, and that our total importations, in the seven years ending December 1843, exceeded twelve millions of lbs.

Our own custom-house returns for the last year and a half, do not exhibit any increase in the importation, a pretty evident sign that the shipments had been already carried to their full extent, which would argue either an enormous decline in the number of sheep, (1,500,000 lbs., on an average of 10 lbs. per fleece, corresponding only to 150,000 head,) or that the consumption in the country is nearly equal to the produce of the flocks kept. The extraordinary rise in the value of this wool, which, at the shipping ports it has been seen, in five years advanced from sixteen to thirty dollars per quintal, has, however, induced the Indians to turn their attention something more to its growth: but it is not expected for some time to come at least, and until the country is more settled, and the facilities of conveyance improved, that Peru will be able to export above 2,000,000 lbs. per annum, after provision has been made for the home consumption—a very small amount when the increasing demand is considered.

According to the respectability of the testimonies which I have received, it may be safely admitted that, at the commencement of the current year, we had received 12,000,000 of lbs. of alpaca wool, all of which was consumed in our manufactures, or shipped to France in the form of yarn, with a corresponding advance upon it. Some few shipments from Liverpool of the raw material have also been made to French and Belgian ports. The greatest share of the spinning and weaving of this article falls to Bradford, where plain and figured stuffs are produced, the warps of which are of cotton and the wefts of alpaca wool, with a beautiful lustre upon them. It is this constant endeavour to devise new and attractive styles of goods, that has fed the market and kept some thousands of our weavers employed, who otherwise must have been dependent upon their parishes for support. In this respect Bradford has been singularly fortunate, where great credit is due to Mr. Titus Salt, through whose intelligence and perseverance the spinning of alpaca wool has been brought to great perfection.

Similar manufactures are carried on at Norwich, as well as other places in England, and have also been extended to Ireland and Scotland. This kind of wool is peculiarly well adapted for the tartan. Owing to its glossiness, it is, to use the trade term, shy of taking dye, in consequence of which a stronger process becomes necessary in order to impregnate the fibres properly with the colouring particle. This difficulty has, however, been equally overcome, and the most delicate colours are now obtained; such as royal blue, scarlet, green, and orange, as seen in the *mouselines-de-laines*, and other ladies' dresses in very general use.* The blacks are superior, and the damask patterns of a showy description. Coat-linings, plain and figured, wear much longer than silk, and under various names are sold as foreign goods. In some instances alpaca takes the place of Angora wool,† and in France has been used both for cashmeres and merinos. It has been proved to be admirably well suited for mixed cotton goods; and so firmly is its reputation now established, that there is every certainty of a growing demand, to meet which an additional quantity will annually be required.

In consequence of the rapid growth of our population, it is universally acknowledged that some great effort must be made to obviate the serious consequences of a surplus, for whom no adequate means of subsistence and

---

* On the Continent, more particularly in Saxony, and other parts of Germany, the object of the flock-master is to grow a fine fleece, without regard to the carcass, whereas with us, as before remarked, it is the reverse. As the sheep increases in weight of flesh, so does the fleece; but, in that case, the wool becomes coarser, and consequently serves only for low coatings, flushings, &c.

† *Vide* Mr. W. Danson's paper on the rise, progress, and extension of alpaca manufactures, read before the Liverpool Polytechnic Society, in Febuary 1842.

* There has been a very animated enquiry for alpaca wools, both black and brown, and some transactions have taken place at a still further advance. The market is still looking up. The goods made from this article are getting much into favour, and are likely to become very fashionable, as her Majesty has lately ordered some patterns of this fabric. Some impovements have also been made in the process of dyeing them; and it is said that the manufacturers are about to bring out some new patterns of a thinner make, adapted for summer wear.—*Liverpool Wool Trade Report for Febuary* 1844.

† The quantity of mohair, or goats' wool, imported into the United Kingdom in 1843, was 575,593 lbs.

employment have been provided. Emigration and colonization are recommended as remedial measures, and to a certain extent would produce the effect desired; but before we send forth our superabundant population to contend with seclusion and other difficulties in a foreign clime, is it not our duty to devise some plan for their support at home?

We cannot extend the surface of our lands; but by rendering those more productive which we possess, we certainly may feed and employ a greater number of individuals than we now do. The lands still left waste and useless in the united kingdom, are allowed to exceed thirty millions of acres, one half of which, by the new draining process and other improved means, may eventually be made available for cultivation; but the remainder, owing to the inequalities of the surface, or other causes, never can be reclaimed. On these, then, let us pasture alpacas. This is an animal whose services come within the scope of our agricultural arrangements, and might also contribute to our amusement; for besides helping to stock our waste lands, which by this means we should render productive, it would become an interesting ornament to the park, where its elegant figure and playful gambols would present a new attraction, in perfect harmony with the most cultivated scenery, although this is not its proper sphere, nor ought it to be placed there until we have a superabundance of stock.

Owing to the greater facilities of supplying distant markets, the rearing of cattle has become fashionable throughout the British isles; and there is scarcely an experiment in rural economy for the increase of live stock that has not been tried, excepting the one here suggested. The naturalization of the alpaca is no longer a matter of theory; nor, indeed, could the prejudices of a few persons, attached to old habits and usages, successfully oppose an undertaking which, if only duly considered, would be found interesting both to the agricultural and manufacturing classes in the empire. Every thing calculated to increase the stock of food, and extend the demand for labour, and open new sources for the supply of raw materials to the several branches of our manufacturing industry, deserves the encouragement of individuals in public as well as in private life.

Such, as already enumerated, would be the benefits gained by the possession of alpaca flocks; while, on the other hand, the expedient would be open to no serious objection—would clash with no existing interest. It would enable those few wealthy and enterprising individuals who have already made experiments on the properties and habits of the animal, to become more extensively useful to their fellow-countrymen; it would become a new and important feature in the history of the agricultural and manufacturing progress made in our country within the present century; and, if only undertaken with spirit and judgment, it would be difficult to assign a limit to the service which such a scheme is calculated to render.

It is the duty of those who are living among us in a state of prosperity, to come to the relief of such persons as are industriously inclined, but cannot find occupation; and this is best done by encouraging enterprises upon which capital and industry may be profitably employed. The unequal state of society requires one class to make sacrifices for the other. Thwarted in their endeavours, and crippled in their circumstances by causes which they can neither control or justly comprehend, our reduced artisans have a right to expect assistance of this kind from the wealthy; and no available remedy for existing distress, even when its operation should be limited, ought to be overlooked. The extension of our markets both at home and abroad, cannot fail to be the result of a more extended use of our waste lands, rendered more important by the production of a new raw material particularly well adapted for our manufactures; and never let it be forgotten that an enlarged demand for goods brings with it increased comforts, and consequent contentment to those who are engaged in producing them.

We cannot give to silks a lasting black colour equal to that stamped upon them in the south of France, and other parts of the Continent. This fact, which our most scientific manufacturers are reluctantly obliged to acknowledge, points out the expediency of using another raw material, in appearance resembling silk, and in quality infinitely more durable, which in some instances does not require the process of dyeing. Black, grey, and brown alpaca wools, for example, when well picked, make a beautiful texture of one uniform colour, which being natural, and consequently better ingrained than could be done by any artificial process, is not affected by acids, or liable to fade. It is on all hands agreed, that there is a deterioration in the condition and prospects of our operatives, arising out of the decreased demand for those articles, the manufacture of which heretofore afforded them a livelihood. Those who have their welfare at heart ought, therefore, to endeavour to see the industry of the country applied to a larger range of objects, which can only be done by our availing ourselves of every new article of growth and manufacture for which our soil, and the habits of our artizans, are adapted.

Breeding of sheep is an occupation well suited to the character and pursuits of our rural population dwelling upon high grounds; so there is now a sufficiency of evidence before the public to show, that far from disturbing the existing relations of society, or interfering with the present distribution of agricultural labour, the adoption of a new stock, such as the one here recommended, is calculated to advance the general, as well as the local interests of the farmer and artisan. This expedient presents a ready and natural resource to the agriculturist employed on poor lands, on those unfitted for cultivation, or thrown out of it by circumstances which he could not prevent—as well as to the artisan who was learned to spin and weave.

The great question with us at the present moment is, how capital can be made to produce the most profitable return. Money, science, and enterprise, we possess beyond every other country. In the history of the civilized world, no parallel can be found of that prosperity which we have attained—of that wealth which we have accumulated; and yet a dark and menacing cloud hangs over the future. Discontent among the lower orders is far from abating; every year the danger arising out of this state of things becomes wider and more alarming; and, if we were to seek a cause for so ominous a change, it might be summed up in these few words—*every branch of industry is overstocked.* And yet it is no one legislative act that could bring relief. This can only be obtained gradually, and by a combination of circumstances; by measures of practical improvement, such as will render the existence of the working classes less precarious, and remove that uneasiness and restlessness prevailing among them.

Why, then, do we neglect the introduction of a new source of employment, when so favourable an opportunity presents itself? The desire to become acquainted with scientific discoveries, as well as with the results of useful experiments, is general among us, and on no two subjects is that desire so strong and so active, as in reference to agriculture and manufactures. Conspicuous as our country is among the other sections of Europe, by

BENEFITS OF NATURALIZATION.

its moral energies and political institutions—by its mechanical and manufacturing skill—by its proficiency in the useful arts; and, however highly cultivated a large portion of our lands may appear, still we have much to learn on the score of agriculture, and on the means of increasing the productions of our soil. In both respects, the naturalization of the alpaca is calculated to promote a part at least of the great and useful purpose in view. Our climate will not allow us to grow silk, neither can we rear the vine, olive, cotton, or tobacco; but by transplanting to a congenial soil so interesting an animal as the one in question, with assiduous care we should not only improve its fleece, together with the manufactures made from it, but also augment the productions of the earth, thus affording a remuneration higher than any derivable from the introduction of other exotics.

The northern division of the kingdom presents an assemblage of lonely mountains, in their wild character resembling the Andes: but, unlike them, covered with useful herbage, which may be seen growing in all the luxuriance of nature, and yet it is mostly left to wither and die neglected. The Cheviot hills contain morasses and peat-bogs, capable of receiving more flocks than those already browsing upon them. The services of the the alpaca, however, ought not to be confined to the Scotch Highlands. The Shetland islands would be an eligible locality for that animal, where it would eat up the vernal grasses and coarse plants growing on the hills, and in the winter feed upon the hay, there called *tehh*, chiefly composed of heaths and rushes cut on the commons. The hardier the breed the better for Shetland. Ireland also has many mountainous tracts, upon which herbage of some kind or other grows to the very summits, and where there is a peculiar tendency of the soil to grass. The bogs of the sister isle, which cover a larger surface than even the mountains, when drained, might equally support alpaca flocks, which would flourish on the wildest and most desolate parts of the Wicklow and Kerry counties, more especially on those tracts belonging to the estates of Lords Wicklow, Fitzwilliam, and Waterford. In the mountainous districts of the north of England, where the blackfaced sheep and Chevoits share between them the land, as well as on Snowden and other hilly tracts in Wales, these animals would equally be an acquisition.

In former times, the kid appeared on the table of the Welsh farmer, and in Spain and Portugal is still considered a delicacy. Owing to the mischievous habits of the goat, it has, however, been nearly driven from the Cambrian hills, where it might be advantageously succeeded by the woolly race from the Andes. There the alpaca would occupy the situations whence the old adventurous inhabitant has been banished. Lastly, several of the middle counties of England comprehend a succession of high and healthy mountains, as well as extended tracts of fenny lands and coarse pasturage, a large portion of which is never trodden by cattle or sheep. We have low and swampy spots upon which neither can be trusted; and yet even there, in the dry season, vegetable substances might be found to feed the new stock. By adopting the plan suggested, the farmer burdened with lands like those above described, would have it in his power to extend his operations, and by this means ultimately render his tenement more profitable to himself and his landlord.

It is estimated that the power at present created by machinery, equals that of 400,000 horses; and calculating that, in food, each horse annually consumes the produce of two acres, it would follow that we have now 800,000 acres of land more to spare than we formerly had. We have, therefore, sufficient room for the admission of the alpaca, without obstructing the progress of our native flocks. Our lands are not only competent to sustain the sheep and cattle already pasturing upon them; but, for reasons previously assigned, by judicious management they may be rendered capable of receiving from six to ten millions of a hardy and abstemious race of wool-bearing animals, with whose beautiful form and figure the public are now familiar, and whose wool is annually rising in estimation. Their introduction would thus become a new and valuable feature in our agricultural and commercial policy—an expedient by means of which we should increase the mass of our exportable articles, and at the same time diffuse a corresponding share of vigour and activity among those who manufacture them. This would be an event calculated to exercise an influence over the destinies of numbers of our fellow-countrymen: for the scheme proposed must lead to results both satisfactory and remunerative, unless it can be shown that we have neither sufficient space, nor sufficient labour, to devote to an enterprise of this kind.

In the actual circumstances of our country, this will scarcely be attempted. On the contrary, the least reflection on the truths expounded in these pages, will render it apparent that we have ample scope for new pastoral undertakings of the nature therein proposed; and we ought not to be so far blinded by prejudice, as not to see the advantages to be derived from the same pursuits as those which placed the ancient Peruvians so much above their neighbours in civilization, and have preserved the modern race from total extinction; for it is by the most intelligent travellers acknowledged, that, if the few Indians living in Bolivia, and on the sierras of Southern Peru, have remained concentrated on the lands of their forefathers, it has chiefly been owing to the attraction of their flocks. An adequate stock of butchers' meat, in like manner as one of corn, sometimes depends upon the caprice of the weather, as well as accidents of the soil; nor will it be easily forgotten that, only fourteen years ago, the rot destroyed upwards of eight millions of sheep in England and Wales.

Our object, avowedly, ought to be, to guard against vicissitudes, by growing whatever else our pastures and climate will allow. This is, therefore, a proposal connected with the various other principles which affect the economical condition of the people; and in the present state of affairs, and looking to the prospect before us, no measures likely to augment the resources of the industrious classes, and help to provide them with the necessaries, and next with the comforts of life, ought to be underrated. The improvements in the mechanical arts, and the greater facilities of conveyance introduced in modern times, have not only quickened and enlarged international intercourse, but also created new wants, both at home and abroad.

The bias of the age tends towards practical knowledge, and the application of scientific powers to the purposes of society. Beginning at home, our great aim therefore ought to be, to increase the capabilities of the country as far as we can; and, to a certain extent, this may be done by the introduction of new flocks. The naturalization of the alpaca in the British isles, is consequently an undertaking which deserves public encouragement; and owing, as we do, so large a share of our commercial prosperity to our woollen manufactures, and intimately connected, as the best interests of our artisans are, with the continuance of that supremacy which those manufactures have attained, it is a matter of surprise that this subject has been so long disregarded.

Britain, at this moment, possesses a much greater

power of producing wealth, and consequently of affording increased employment to the labouring classes, than she did half a century ago, if her resources are only fairly brought into action. With this view, let us then begin by growing an article suited to our soil and climate—an article which, in the end, will prove beneficial to the manufacturer, and besides render six or eight millions of acres of our useless lands productive. The increasing tendency of continental politics to exclude us from those markets which we have been accustomed to supply, calls for corresponding exertions on our part. The rapid strides made by our neighbours in several branches of manufacture; the capital invested by them in spinning and weaving cotton, silk, and wool; as well as the association of German customs, rapidly rising into manufacturing power and political influence—render it imperative upon us not to neglect any available means of invigorating our own commerce.

The onward pressure of events obliges our rulers to keep pace with other nations in the arts, sciences, and manufactures; and, after what has been here stated, it would be superfluous to point out the necessity of making up any deficiency resulting from the deterioration in home-grown wools, or the diminished consumption of British woollens abroad. Fashions, besides, wear out, taste changes, and new connexions are formed by the consumer. In the stirring age in which we live, nations, like individuals, must compete with those who seek to outstrip them; and without entering into invidious comparisons, or indulging in the smallest ebullition of national feeling, it will readily be admitted, that, in order to keep his ground in the foreign market, the manufacturer must vary his goods, and adapt them to the prevailing taste, besides increasing the number of articles which he sends thither for sale.

CHAPTER V.

RESULTS OF THE EXPERIMENTS ALREADY MADE TO NATURALIZE THE ALPACA.

FORTUNATELY, within the last few years, attention has been directed towards the scheme of naturalizing the alpaca; and we have among us individuals who may justly claim the credit of having done all in their power to promote an enquiry into the subject, so as to cause its merits to be fully investigated.* Our country presents splendid examples of the great good done through the exertions of private persons, and certainly here there is

* In 1841, the Highland and Agricultural Society offered a gold medal for a "satisfactory account, founded on actual observation and experiment, to naturalize in Scotland the alpaca." Although the writer of these pages did not possess the precise kind of information sought for in the preceding notice, still, as the project was yet in its infancy, and, owing to the deficiency of stock, could not at that time have afforded results calculated to guide others, he nevertheless ventured to compete; and at a meeting, held in January 1842, the premium was awarded to him. With permission of the Society, some portions of his essay, then presented, are embodied in the present work. From their programme of prizes it appears, that this year the Highland and Agricultural Society have offered premiums for the best pair of alpacas born and bred in this country, and also for the best two, male and female, imported. It is to be hoped that the judges will look carefully into the pedigrees of both classes, in order to ascertain that they are of the really pure breed, and do not belong to the spurious race, obtained by crossing.

a fair opportunity of adding to the number.* Several of our nobility and gentry, on a small scale, have already made the trial, although, as it will hereafter appear, by no means in such a manner as to give the animal a fair chance. A few little flocks, however, for some years past have been kept in various parts of the kingdom, and towards the close of 1840, it was estimated that we then had among us from fifty to seventy Andes sheep, principally llamas and the mixed breed, the proportion of pure alpacas being very small, and the vicunas not exceeding five.

As a wool-bearing animal, the value of the alpaca at that time was scarcely known. All the breeds were treated as mere curiosities, and kept in gentlemen's parks and paddocks, or belonged to the owners of zoological gardens or travelling shows. Nobody thought of breeding the alpaca as a farm-stock, capable hereafter of peopling our hills and serving as the basis of a new manufacture. Although the few possessors of the pure breed had the evidence before their eyes of a heavy and beautifully fine fleece; although they saw their wives and daughters clad in the glossy textures made from this kind of wool, and must have been sensible that, coming from a foreign land, the supply would always be scanty and precarious; still no one seemed to recollect that it was the commerce of our merchants, and the ingenuity of our artisans, which laid the foundation of our national wealth, and had made us known in the remotest quarters of the globe. No one appeared struck with the idea, that, if the manufacturer furnishes the merchant with goods, somebody must provide the spinner and weaver with the raw material required to make them; and that it is better that this should be done by the farmer at home, than expose the progress of our looms to contingencies.

Within the last three years, whalers and trading vessels from the South Seas, have occasionally brought over a few alpacas, or animals sold to the captains as such; but, there is reason to believe, often injudiciously selected. These animals have chiefly arrived at Liverpool as presents, or as the captain's adventure; and being considered objects of natural history, and also few in number, it has not been usual to report them at the custom-house. It would, therefore, be impossible to state the number actually arrived with any thing like accuracy, and more so to class and trace them. From the enquiries which I have been able to institute, I am, however, inclined to think, that, within the period above specified, about 200 have been landed at our several ports, mostly in a woful plight. Of these, it can scarcely be thought that more than 150 survived; and of that number, at least twenty-five have passed over to the Continent. Calculating, then, that about the same number have, in the interval, been born in the kingdom, this would leave our total stock of pure and mixed breeds at 210, divided perhaps among thirty individuals, and located in various parts of the country.

This, I must confess, is a much smaller amount than I anticipated when I wrote my second treatise at the commencement of 1841; but had it not been for the most unfortunate accident—the most melancholy occur-

* Of the great service that may be rendered to a country by the exertions of even one rich and spirited individual, France affords a brilliant example. The late M. Ternaux, the well known manufacturer and deputy, impressed with a due sense of the advantages which would accrue to his native land from the possession of the Cashmere goat, at his own expense sent out a special agent, who purchased 1800 of these animals for him, only one half of which arrived safe. Some were sent to the Pyrenees, others to Versailles, and a few sold to amateurs. By this distribution, it is evident that he was desirous of ascertaining which climate and position in France suited them best.

RESULTS OF NATURALIZATION.

rence that possibly can be imagined—in the early part of last year the above number would have received an important augmentation. The captain of the Sir Charles Napier, bound from Islay to Liverpool, was entrusted with the embarkation of 254 female alpacas, purposely selected in the last stage of pregnancy, and 20 males. Unaware of the consequences, he filled the lower hold of his vessel with guano manure, placing the live stock, for whose reception every possible preparation had in other respects been made, in the 'tween-decks. When out at sea, the guano heated, and the effluvia, loaded with ammoniacal, or other strongly deleterious properties, rising gradually, suffocated the poor animals, but not till the greater part of the females had given birth to their young, some of them prematurely. When the vessel was reported at the Liverpool custom-house, on the 15th of April 1843, four only were left alive. Thus, through an act of inadvertency, perished the first cargo of these interesting animals ever attempted to be brought to our shores.

As a remarkable instance of the vicissitudes to which these animals have been subjected, when once taken from their native mountains, it may be mentioned that, two years ago, four alpacas, brought to Liverpool, round Cape Horn, were there purchased on speculation at the low price of £75, and shipped for the Cape of Good Hope. They arrived in good health, and were advertised for sale; but no one offering to take them at the advance required, they were sent up the country and located on a farm in the Swallendaur district, where, according to the last accounts, they were doing well. The new and increasing trade, lately opened between Australia and the western shores of America, has also enabled some amateurs in the former to procure a few alpacas direct from Peru; but the experiment is of too recent a date to afford precise results. Some hilly portions of the Australian colonies are, no doubt, suited to their growth, and thither they may be carried at a much cheaper rate than to Europe.

Sensible of the value of their acquisition in obtaining from Spain the merino breed of sheep, there is, at the present moment in Germany, a strong desire to form alpaca flocks. The king of Bavaria has evinced the deep interest which he feels upon this subject, and through the Royal Academy of Sciences at Munich, lately addressed to the Agricultural Society of England the following queries. "1st, How many alpacas have been imported into England? 2d. What is the amount of their wool sent over annually from abroad? 3d. Is the wool preferable to that of common sheep? 4th. Can the animal be easily acclimatized and fed in mountain districts? 5th. Have any experiments been made in England with alpaca wool? 6th. Does the wool admit of being spun by the spinning-wheels in use for the wool of common sheep, or does it require a particular sort of wheel? 7th. Has the importation of alpacas proved useful and beneficial, and what is the general opinion of the wool for manufacturing purposes?" The King of Prussia has also directed inquiries to be made respecting the habits and properties of the animal, as well as to ascertain the results of such breeding experiments, made by our agriculturists, as can be authenticated.

The great efforts, made and making by both monarchs, for the advancement of agriculture among their subjects—a theme of general admiration with those associations and individuals at present so laudably engaged in endeavouring to augment the productions of our own soil—warrant the belief that sovereigns, so spirited and patriotic as they are, will lose no opportunity of conferring this additional boon upon their subjects.* The late King of Holland also expressed a wish to have alpaca flocks. A small number have found their way into France; but it does not appear that any one there has hitherto taken an interest in their propagation as farm stock.

My present effort would, however, be incomplete, and indeed my motives for again appearing before the public on the same subject might be mistaken, it if were not in my power to show that I have used all due diligence in obtaining the actual results of such trials, with this kind of sheep, as have already been made in the several parts of our country. With this view, I wrote to the principal owners whose names I could procure, frankly avowing my object, and requesting particulars. With few exceptions I was favoured with a reply; but as it is evident that, in most instances, the experiments were commenced under a wrong impression, and pursued in a manner by no means calculated to establish a proper method for the treatment of large flocks, I deem it best to subjoin each report, as near as I can in its original form, reserving to myself the right of afterwards making such remarks upon the facts and opinions therein set forth, as circumstances may warrant.

It was through the opportunity of contemplating the beauty, and studying the properties of Mr. Cross's (late of the Surrey Zoological Gardens) white and brown female alpaca, exhibited from 1810 to 1816, that the British public first became acquainted with the value of this interesting race of quadrupeds.† This specimen was originally brought from Lima, where it had been a pet; and the perforations in its ears, in which ornamental rings had been placed, were still visible. Its graceful attitudes, gentle disposition, and playful manners, were particularly attractive. Ladies frequently caressed it as if it had been a child. Although kept in the unwholesome atmosphere of a crowded city, pent up in a close room, and unavoidably fed on unsuitable diet, it nevertheless attained the usual age; thus affording as satisfactory an example of hardihood as could be wished. But let Mr. Cross state the result of his own experience himself, he having kindly furnished me with the following, dated May 3:—

"As you are desirous of having the result of my observations relative to the alpacas which have fallen under my notice, I beg to say that the first one brought to this country came into my possession after being for two years in that of Mr. De Tastet of Halshead, in Essex, who exchanged it with —— Tharpe, Esq. of Chippenham Park, near Newmarket, for a pair of magnificent coach-horses. The latter gentleman kept it for about two years, in the hope that one might arrive from Peru of a different sex, as he was anxious to breed from them in consequence of the fineness of their wool. Disappointed in his object, I bought it of him for one hundred guineas, and exhibited it about six years; consequently it must have been eleven or twelve years old when it died.

"It was fonder of browsing than grazing. One remarkable fact I cannot help mentioning, and that is,

---

* The King of Prussia has a model farm, and agricultural colleges have also been established in various parts of his kingdom. For a similar purpose, the King of Bavaria has given up one of his own palaces. So much do both monarchs consider the improvement of the soil as a national object, that, in order to promote it, occasionally they sanction grants of money out of the public treasury for any new experiment properly recommended to their notice.
† The two wood-cuts accompanying this work are from drawings of the animal in Mr. Cross's possession.

RESULTS OF NATURALIZATION.

that it never drank any thing during the whole time I had it, though repeatedly offered drink. I fed it upon bran, oats, carrots, and hay; occasionally in the season a little green tares. Its wool was about eighteen inches long, mixed with some trifling portions of hair.* It was remarkably tame, and I may say affectionate. I have since had several others of various colours, some quite black, and others piebald, &c. Having had them so often, I let them take their chance in the stalls, giving them the opportunity of running into a paddock, and they always did well. The last pair I had were perfectly black, and I sold them to Mr. Advenant, who immediately took them over to the King of Bavaria. The late King of Wirtemberg, and other continental monarhs, also had some of me. I think they might be introduced into some parts of this country with great advantage, particularly in the hilly parts of Scotland and Ireland, where they could have an opportunity of browsing as well as grazing. Occasionally they will breed with the llama; and a more elegant animal than the offspring cannot be imagined, but whether the latter will breed again, I cannot say."

About the time Mr. Cross was exhibiting his interesting specimen, the late Duchess of York had four or five llama and alpaca pets at Oatlands, where she took great delight in watching their sportive antics on the lawn, or contemplating their intelligent and expressive countenances, greatly resembling that of the gazelle. They ran the chance of all exotics, whether animal or vegetable; left to the care of servants who, when the master and mistress are away, usually treat both as mere matters of course, and often with a strong feeling of prejudice. When the Duchess died, these pets necessarily were dispersed; and, in all probability, at that early period breeding with them was deemed an impracticability. They, however, lived long enough at Oatlands to render it apparent that they are of a hardy race, although the old and plain-spoken park-keeper has more than once been heard to say, that, while under his charge, they were not in their proper element, the grass being too *firm*—meaning too rich and good.

In 1817, the late D. Bennet, Esq. of Farringdon House, Berks, received a pair of alpacas, and fed them, as he did his sheep, with hay and turnips in the winter. He found them hardy and healthy; and noticed that they required little care. From this stock he reared fifteen, of which the greatest number he had in his possession at one time was eight. Generally the young ones passed into other hands. The present number at Farringdon is, I believe, five.

From Viscount Ingestre I was favoured with the following, under date of May 31st:—"In answer to the queries which you put to me, I beg to state, 1st, That the animals I imported were a pair of alpacas, and that I shipped them from Valparaiso in the latter end of the year 1825. 2ndly, That they stood the voyage remarkably well. 3dly, The female had, three or four times, one young one at a time. And 4thly, They were for some three or four years at Earl Talbot's, at Ingestre, to whom I gave them, and who afterwards, I believe, made a present of the whole stock to the Zoological Society in Regent's Park. I will add that I had them shorn once or twice, and had the wool spun, which made a cloth of the softest texture possible. I have no doubt that they might be naturalized in this country," &c.

* The circumstance of a few hairs being intermixed with the fleece, was, I do not doubt, attributable to the age of the animal, and its having retained the same coat upon its back for several years. If it was four or five years old when it left its native shore, its age must have been even greater than that estimated by Mr. Cross.

The next person who seems to have taken a fancy to these animals, was Thomas Stevenson, Esq. of Oban, Argyleshire, who, under date of the 15th of last March, politely forwarded to me the annexed report: " Fourteen years ago, a son of mine in Peru, shipped on board of a merchantman a dozen alpacas for me, with an understanding with the commander, that he was to receive for payment of freight one half of whatever number might arrive safe in England. The object of this arrangement was to induce the captain to take greater care of them than he otherwise would do; yet of the dozen only four reached Liverpool, and of course I only got two, a male and female, which were about a year and nine months old when they reached Oban. Although I had been long in South America, I had never seen an alpaca, and was therefore ignorant of the proper mode of treating them; so I fed them in the same way as we do Highland cattle, and found it to answer remarkably well, in so far as their health and growth where concerned. They were driven out with my milch cows to pasture summer and winter. During the night they got a little hay or straw: and in winter, when snow covered the ground, a little corn in the sheaf was placed before them. They were fond of all vegetables and shrubs; particularly so of hedges and the tops of young trees. I never weighed them; but I should think they would have weighed from eleven to twelve stones of sixteen pounds to the stone. The male was very strong, and I have seen him canter easily with a stout man on his back. Their wool was very fine; but I made no use of it further than manufacturing some of it into stockings for my family. I am sorry to say that they never bred.

"A year afterwards, my son shipped a dozen llamas for me; but I only got a pair, eight having died during the voyage, and the commander of the vessel reserved the other two for himself. My two received the very same treatment as I had observed towards the alpacas; and, when two years old, the female had a young one, and continued to have one regularly every year about the month of April. She went a year with young. Of the offspring about one half lived; of which, the females began to bear when two years old. I sold my young stock to various persons; and two years ago, finding that I had only one female left, and that the old one, I sold my whole stock, which consisted of five, having five years previously sold my alpacas.

"The alpacas and llamas lived very quietly together until the latter had their first young one, when the male alpaca became extremely jealous and furious, and on several occasions leapt a wall five feet high, and broke through a dozen men, to beat the male llama, which, being slightly the heavier of the two, he did, carrying his anger so far as even to beat the young one when he could reach him. Besides the alpacas and llamas above mentioned, two years and a half ago I had another Peruvian animal, called the vicuna, generally considered to be of the same species. It is not, however, so large, being about the size of a fallow-deer, but infinitely more graceful and beautiful. This animal runs wild in Peru, and I could never tame mine, although I received it when very young. It was led out every morning to a small inclosed park. The wool of the vicuna exceeded in fineness any wool I have ever seen. I tried to have some of it manufactured into a shawl at Glasgow, but could not succeed, and at last I lost my packet of it in transmitting it from one manufacturer to another. I was so unfortunate as to lose this beautiful animal by a boy striking it on the heart with a stone, which caused instant death. It was a female; and what made me lament my loss the more, was the circumstance, that the poor creature was six months gone with young to the llama. My son a second time sent me eight alpacas, but they all died on their passage over.

To subsequent enquiries Mr. Stevenson, under date of March 30th, furnished me with the following additional particulars:—" The grounds upon which I fed my alpacas and llamas were of different descriptions, being partly hill and partly plain, and they seemed to agree equally well with them. In my former letter, I forgot to state that, during the warm months of June, July, and August, my alpacas and llamas were left in the fields all night. The total number of births I had was, I think, eleven. Of these, six came to their full growth, one was killed by accident, and four died when a few days old. The greatest care is required to be taken of them till they are two or three weeks old, after which there is no fear of them. The mothers are very fond of their young, and take great care of them, spitting at any one who comes near them. I did not try to cross the alpaca and llama; indeed, I never thought of it. I regret to hear that you find so much difficulty in awakening the attention of farmers to this subject. I have just had a letter from Lady ——, making enquiries for a friend in Germany; and I should not wonder if the Germans do not forestall us in the acquisition of alpacas. I have heard from my son in Peru, that he has lately, on two occasions, shipped for me four alpacas, but they all died on the voyage. I think the way you propose bringing them over will be a great improvement."

From the same party, under date of the 10th of April, I was favoured with the following:—"I have delayed until now acknowledging the receipt of your last esteemed favour, in order that, before doing so, I might have an opportunity of carefully reading your very interesting work on the naturalization of the alpaca. I have indeed had much pleasure in perusing it, and I only wonder it has not been the means of causing farmers to take a greater interest in the matter than they at present seem to do.

"At p. 14 of your book, you say that the llama goes seven months with young. I may mention, that those I had went between eleven and twelve months. The female was invariably covered by the male two ro three days after she had a lamb, and, from the singularity of this circumstance, it attracted my very particular attention, and I regularly marked down the date on which the female was covered, and found that she went with lamb a very few days less than a year. I may add, that I have in my possession a stone representation of a llama at rest, as you mention in p. 16, which was taken out of an Indian grave in Peru. I may also state, that I have two grandsons with me from Tacna, who remember having often seen the dried flesh of the llama. They agree with you in saying, that the Indians are very fond of eating it."

By Robert Gill, Esq., I was favoured with the subjoined, dated Manchester, April 15:—"It is quite true that I was one of the first who possessed the llama and alpaca. I also succeeded in breeding them, I think, to the extent of three; but, as I have no memoranda, I can only speak from memory. I also possessed two vicunas, but was not successful in breeding from them; nor have I heard that any one here succeeded in breeding them. In consequence of my changing my residence about five years ago, I parted with the few animals of this class which I then possessed. I am glad to find you are taking up so interesting a subject. I had not heard of your work; but shall have much pleasure in reading it, as well as the forthcoming one. Should you succeed in adding to the number of our domestic animals, you will deserve the thanks of the country at large. Much remains to be done, both in quadrupeds and the feathered race."

Joseph Hegan, Esq. of Liverpool, after stating that he was the person who presented to the Earl of Derby the first alpacas his lordship had, about five years ago, under date of April 20, writes thus:—"For two years I lived at Arrow Hall, Cheshire, and while there had three or four of these animals. The survivor of these, with the progeny of one female, have been for the last two years on a farm in Ireland, belonging to Mr. W. Danson of this town, and I really know nothing of them. The manager of the farm is Mr. Bell of Gunsboro,' near Listowel. I am sure he will readily give you all the information in his power; and he has now had a fair opportunity of ascertaining the habits of the animal, from its birth onwards. Those under the charge of Mr. Bell are the pure breed—unmixed alpacas."

Charles Tayleure, Esq. of Parkfield, near Liverpool, in a note, dated April 11, speaks thus:—"In reply to your enquiries, I beg to say, that my memory is not sufficiently good to enable me to state in what year I imported the first alpacas and vicunas. I recollect that there were a pair of each, and that the alpacas had a young one, the others none. I had the misfortune to have one of the vicunas killed by a dog; and in consequence of some subsequent importations of alpacas being diseased, and the disease spreading to the others, I put them under the charge of a shepherd in the neighbourhood, who, by administering too strong medicines, killed the greater part of them. This tended to disgust me; and, not long after, the only two alpacas that I had left I gave to Lord Derby. On a really mountainous country I consider that they would do well. Cold does not affect them, but diet does."

From Henry Lees Edwards, Esq. of Pyenest, near Halifax, under date April 6, I received the annexed:—"In reply to your enquiries I beg to state, that in 1839 I imported from Valparaiso six alpacas and two vicunas. In 1841, of six alpacas shipped for me, only two arrived safe, and in the same year I purchased three in Liverpool, making a total of eleven alpacas and two vicunas. Of the former eight have died, leaving my present stock three alpacas and two vicunas. From the latter I have had no lambs, but from eight alpacas in field I had eight lambs, chiefly premature births, only two of which lived to twelve months, and them I have also lost. These animals have been much subject to scab, which is difficult to remove from them, and mine were seldom free. They were kept in a good grass field on the side of a hill, a dry pasture, but not short grass like the hill tops. In the beginning they had a good deal of hard food—oats, beans, &c., besides grass and hay—but when they died so rapidly, I discontinued hard food, and now only give them grass, hay, and vegetables."

The Earl of Derby's being mixed up with the alpaca question, seems to have been purely accidental. Pursuing that refined taste for rare objects of natural history which has always distinguished his lordship, he made the acquisition of a few llamas, and added them to his splendid menagerie at Knowsley. There they were seen by Mr. Hegan, who happened to mention that he had some alpacas on his property in Cheshire, of which two were eventually transferred to his lordship, who subsequently obtained as many more from Mr. Tayleure of Liverpool. Treating both varieties as mere curiosities, and seemingly never intending to use the preferable one as farm stock, his lordship allowed them to cross, and the result was births in the ordinary course. Both llamas and alpacas, kept separately, also bred almost every

RESULTS OF NATURALIZATION.

year; but, besides these, his lordship procured a pair of guanacoes, such at least they were called by the seller, although I am inclined to think most erroneously.

At the request of some gentlemen in Liverpool, his lordship allowed several specimens of his flock to be exhibited at the Royal Agricultural Society's meeting held there in July 1842, as already mentioned, and the menagerie has always been open to visitors; but his lordship has not, as yet, considered these animals sufficiently in the light of stock to be placed under the care of a farm bailiff. His lordship, indeed, seems only to have viewed them in the character of a *virtuoso*, and with no ulterior object; but as his lordship has been kind enough to favour me with an opinion, under date of April 5, I consider it my duty to annex it, in his lordship's own words:—

"I am sorry to say that the result of my experience has rendered me less sanguine than I have been, of these animals becoming so acclimated as to form a very valuable addition to our general farming stock in this country, though, as a matter of curiosity, I have no doubt whatever as to their success. I do not doubt their hardihood, and their being easily fed; but they will not, I fear, answer to form large flocks like sheep, unless it be solely of females, as the males do not agree well together. The disorder they are subject to is also infectious, and requires too much trouble and watchfulness to get rid of which must always tend to diminish their value to the farmer, notwithstanding the value of their fleece, which, instead of being deteriorated, I am led to believe, is rather improved by our climate, or the care taken of the animals." His lordship adds, that "he had never been able to obtain more than one specimen of the vicuna, and that a female."

On the 22d of February the Marquis of Breadalbane condescendingly informed me that "he had a few alpacas for a short time, but they all died with the exception of one;" adding, "that it was the opinion of his people who had charge of them, that the pasturage was too rich, and that they would have done better on hill ground."

Understanding that the illustrious consort of our patriotic Queen had so far identified himself with the fortunes of the British farmer as to purchase a pair of alpacas, with the intention of allowing them the range of his grounds, I addressed a note to G. E. Anson, Esq., his royal highness's treasurer, and, under date of March 5th, was honoured with a reply in these words—"It is true that there are two alpacas at Windsor, but, as yet, no use has been made of them." I have since received a sample of black wool, clipped from the Prince's male alpaca, and full ten inches long. In quality it is much superior to the imported, being exceedingly soft and moist to the touch—a proof that the fibres contain more yolk, or, in other words, the animal has drawn more appropriate nourishment from our pastures than it could on the Peruvian mountains. The wool of the female is shorter, but equally fine and lustrous.*

Under date of May 31st, I was favoured with the annexed from A. G. Stirling, Esq. of Craigbarnet Place, Lennoxtown, near Glasgow:—"I received your letter concerning the alpaca, &c.—It had for several years been my wish to procure a couple of these animals from their native mountains; but, after various attempts, I found I could not succeed. Latterly, however, I was fortunate enough to obtain a male and female from the Earl of Derby's stock, which were sent here about the latter end of last August. My motive was—*first*, that, considering the great altitude at which alpacas live at home, and the sort of food they subsist upon, I thought that our hilly bent in Scotland, which neither sheep, cow, nor horse will eat, might prove well adapted for them, taking into account the coarser herbage upon which they thrive; *secondly*, I wished to keep these Peruvians in my sheep-park with the sheep, in front of this house, so that I might be enabled to form an opinion of their habits, &c.

"Now, sir, from ten months' observation I am enabled to state, that I have found them *most docile*, mild in temper, and never attempting to break a fence, and agreeing perfectly well with the sheep and young lambs. During our last winter, which was more severe than usual in this country, with many vicissitudes as to climate, and attended with weeks of deep snow, I thought it necessary to commence giving them some food, and began with rye-grass-hay and turnip. After some little time the keeper told me that they were not eating the rye-grass-hay so well as at first. I then desired that a coarser kind of meadow hay should be given to them. This they greedily ate, and left the other. They next seemed to tire of the turnip, and I ordered each of them to have a handful of oats, which they eat freely of at first. However, in about three weeks they also became indifferent to the oats. A handful of beans was next tried; they fell to them most greedily, and never lost their relish for them. By way of experiment, I desired the keeper to mix the oats and beans together; and, as a proof of their partiality for the beans, it may be stated, that they picked the latter out and left the oats; which, if they are to be winter-fed, evinces that meadow hay and beans is the food they like and would thrive upon.

"A small shed had been put up in the park for them, to which they generally resorted at night; but, when the snow was at the deepest, and the wind blowing hard and piercing, these animals left their shed, and picked up what grass they could get at the roots of trees. This shows that they are impervious to cold. Not so the sheep, for they were then cowering down under shelter, wherever they could find it; which proves to me that our climate would agree well with alpacas, and that they would exist where our sheep would die. We hope that the female is with young, which time will show. Both male and female are jet black, and there is a small speck of white, about the size of a shilling, upon the nose of the male. In so far as my experience goes, I can safely say that they have never had a day's illness since they came here. Their clip of last year amounted to 17½ lbs., which are beautiful silky fleeces, and which said silky wool is still in my possession."

Under date of June 6th, the same gentleman favoured me with the annexed:—"I omitted to state that I have read your able treatise concerning the alpaca, and which I did with great satisfaction. I have often referred to it as being the best and clearest: and, as I go entirely along with you in your views, I can have no doubt that, when the subject is better understood, the animal itself better known, and a more expeditious method contrived to bring them to Britain, we shall have thousands of them. When known, their docility, their temperate habits, their hardiness, and, I may add, their easy keep, will erelong bring them into general notice. I can an-

---

* Since the foregoing was in type, I have received information that the male alpaca died at Windsor in the middle of last month, from no apparent cause. I have not been able to ascertain whether any dissection was made of the body, but I very much fear the animal had been overfed. It was a great pet with the Queen; and the weight of its last fleece equalled 16 1-2 lbs., some of it eleven inches long, softer, and with a more wavy appearance than any imported. It is understood that her Majesty proposes to have textures made of it for her own use. It is hoped that the bereft female is in lamb, and no doubt a mate will be sought out for her.

RESULTS OF NATURALIZATION.

ewer, without the fear of being contradicted, that they will thrive and breed in Scotland, equal, if not superior, to our native blackfaced sheep.

"Since I wrote you, I am sure that you will be glad to learn that, on the 1st inst., a little female alpaca made its appearance. It is doing quite well, and a very pretty creature it is—jet black, beautifully shaped, and extremely lively, the very picture of its mamma and papa, both of whom are most kind to it. We have had constant rain, notwithstanding which the little alpaca seems to defy the weather. I do not know if there was ever a young alpaca dropped in Scotland before—certainly not in this county—I mean upon a *fair trial;* for those I possess have had no particular favour shown them, and went through every vicissitude of weather, equally if not with less trouble, than our own native sheep; and during the whole time they have been in my possession, never had an hour's illness."

The introduction of the alpaca into the mountains of Ireland is a new and important feature in the experiment. The attempt was first made by Robert Bell, Esq. of Villa House, near Listowel, in the county of Kerry, a practical and observant farmer, who seems to have started with the determination of treating the animals, as nearly as he could, with the same fare, and in the same manner, as they are on their native hills—that is, no petting, and plenty of exposure. This is by far the most judicious plan, and it is to be regretted that it has not been adopted earlier. Subjoined, is a report addressed to me by that gentleman, under date of March 18:—

"Agreeably to your request, I have much pleasure in giving you a brief statement relative to the small herd of alpacas which we have here. These beautiful and interesting animals arrived at this place in the summer of 1842, previous to which, after being landed at Liverpool from their native mountains in Peru, they were kept in that neighbourhood for some time. They are the only animals of this kind ever introduced into this country, and have been admired by numbers of persons, many of whom came very long distances to see them. Of course, it is quite unnecessary to give you a description of these graceful creatures; for any one who has perused your publication on the 'Naturalization of the Alpaca,' and has seen and observed the habits of the animal in question, will readily perceive how thoroughly you are acquainted with the subject of your work in its minutest details; but I may inform you, that the alpacas on this farm are of various colours, some being brown, others black, and one perfectly white. They have not been shorn since the month of June 1841, and the average length of their wool at this time is eleven inches, and so firm to their bodies, that the smallest lock cannot be pulled off without great force, therefore they never lose a bit. It is exceedingly fine and silky; indeed, very much finer than any alpaca wool I have yet seen imported into England; and, during the two years they have been here, there is a visible improvement in the texture of their coat, and I think that the wool of the alpaca lamb here is superior in fineness even to that of the vicuna. I have frequently examined them very closely, but could never find upon them a wool-tick, or any vermin whatever, to which ordinary sheep are subject; and I was very much afraid, during the exceedingly hot weather which we had last summer, that, from their great weight of fleece, they would be attacked by the fly; but I am glad to say that no such casualty befell them, although sheep contiguous to their pasture, were much injured by it. I have never, even after a whole day's rain, found them wet to the skin, for their wool on becoming wet on the outside or surface, mats together, and becomes quite impervious to the heaviest showers. I certainly do not exaggerate when I say, that each of the old alpacas here would clip at this time upwards of thirty pounds of wool.

"I find they usually copulate here in the month of October or November, although the female invariably takes the ram immediately after having brought forth her young, which is generally in the month of May or June. At the age of nine months the produce of the feminine gender will begin to breed, at which time their wool will be found to be six inches long, and their height to the shoulder thirty-four, to the top of the head, fifty-one inches. The size of a full-grown alpaca is, to the shoulder, thirty-seven, and to the top of the head, fifty-nine inches. You state, in your publication, that they do not begin to breed till three years old, and I have no doubt you are quite correct as regards those reared on the Andes, or other high and barren mountains; but when brought to a more genial climate and better pasture, the case becomes different, and they will breed, if allowed, at the age I have stated. It is the same way with cattle and sheep, when reared on a mountain and scantily fed, they will not breed till they are three, and sometimes four years old, but, when reared on good pasture, they will begin to breed the first year. An alpaca lamb, before being weaned, is always very fat; and I think, at this age, they would be most excellent food; but, when they have attained their first year or so, like most fine wool-growing animals, they do not take on much flesh, no matter how good their pasture. They have no propensity to stray, or wander away from their accustomed feeding-ground, and the lowest fence will be found sufficient to keep them in the field. I have never attempted to obtain a cross with them and the common sheep, nor do I think such a result desirable, even if it could be obtained, of which I have my doubts.

"The alpacas are exceedingly playful, and, to see them to full perfection, a dog should be taken into the field beside them; and, as they run at and play with the dog, their fine and noble positions are displayed to most advantage. They are very sagacious animals. I had a young one, whose mother died when lambing on her way from Liverpool here, that would answer to its name, and follow me about the farm like a dog. From what I have observed of the nature and habits of the alpaca, I do most heartily confirm your statement, 'that they would live where a sheep would starve,' and would be most valuable as a breeding stock in the United Kingdom. They are peculiarly well adapted to mountainous districts, however coarse the herbage, if the ground be dry; although, at the same time, I will say that the alpaca is as fond of a bite of good sweet grass as any animal I know of. The herd of alpacas here are the property of J. J. Hegan, Esq. of Liverpool, who was induced to send them to Ireland at the request of William Danson, Esq., whose praiseworthy exertions to introduce them extensively into the United Kingdom are well known. If your knowledge of South America, and the districts where the alpacas most abound, suggests to you a route by which they can be brought to our shores at a greatly diminished cost, compared with that of those which have hitherto arrived, you will, by making it known, confer an eventual benefit upon the country generally; for, undoubtedly, we should then soon have it in our power to establish this profitable and desirable breed of animals, permanently and extensively, in our isles."

In the middle of May I visited Gunnersbury Park, near Acton, where Baron Rothschild had just received a small black alpaca, which in the daytime was tethered on the lawn, and in the evening shut up in a stable by itself. The Baron, I am given to understand, is much pleased with the animal, and has expressed a wish to form a little flock. It is to be hoped that he will; for most assuredly, he could not evince a better taste, or lay out a trifling portion of his splendid fortune to more advantage, than by patronizing so useful a scheme as the one here unfolded. In the zoological gardens, now so very generally kept in various parts of the kingdom, llamas and alpacas, or mixed breeds, may be seen; but I did not consider that, in those situations I was called upon to notice them.

CHAPTER VI.

REMARKS ON THE SAME—ERRORS COMMITTED BY OUR EARLY BREEDERS.

In perusing the several reports and statements above inserted, two particular facts strike the reader; viz. the great difficulty experienced in obtaining alpacas, and when obtained, the unwise—nay, in repeated instances, the unnatural—use made of them. Scarcely of one could the age or pedigree be traced by the eventual purchaser. With the exception of Mr. Hegan, no one of the early breeders seems to have been sensible of the importance of keeping the stock pure; and hence those who could blended one variety with the other, as if numbers had been their only aim, unaware that, for reasons previously explained, in the course of time they must have been left entirely without stock.

The llama, it would seem, was the first and more generally used for breeding, the offspring regularly passing into the hands of some travelling showman, or keeper of zoological gardens, under the name of short-coated, and subsequently the cross under that of long-coated llamas, the pair frequently selling as high as £150. No distinction appears to have been made between the two varieties. Almost, one would think, that the llama had the preference; and, as soon as alpacas became fashionable, some of these very same mixed natives found their way into gentlemen's parks, when, after a three or four years' experiment, the genuine race was condemned, because these hybrids did not breed.

As a wool-bearing animal, the owners, nevertheless, had in the alpaca all that could be wished—why, then, degenerate the breed by a mixture with an inferior race? This is not the way our farmers treat either their flocks or their herds. As a beast of burden, the llama was of no use to us. It could only serve as a rarity for an exhibition, where it was called a guanaco as often as by any other name. In other respects it is valueless, and it is a pity that so much money has been expended in bringing over and rearing this variety, when the useful one might have been had and propagated at the same expense. As regards the vicuna, it has already been shown that the Incas, in the zenith of their power, and with infinitely greater facilities than we possess, never succeeded in its domestication; but as for the guanaco, I very much doubt whether one was ever yet seen in our country.

This animal, as Mr. Darwin affirms, certainly may, and has, in some solitary instances, been tamed; and among the extreme southern tribes, is occasionally noticed as a pet near the hunter's cabin; but the Indian of the central Andes, who is blessed with his llama and alpaca, would never burden himself with such an encumbrance. The first time he and his family were pressed with hunger, the best joints would find their way into his *olla*. Some naturalist, as a curiosity, or as was done in Josephine Bonaparte's time, might be disposed purposely to procure and bring over one, caught young;* but that a trading captain, either at Arica or Islay, should feel disposed to purchase a guanaco when the tame breeds are within his reach, is out of the question. As stock, that animal could be of no use, unless it was wished to introduce the bezoar-stone again into medicine, or vary our venison.

A scientific officer belonging to the Antarctic Expedition under Captain Ross, writing from the Falkland Islands, where the Erebus and Terror put in for the purpose of repairing and making observations, at the time wrote to a friend as follows:—" I have tamed a guanaco from Patagonia. He lies down before the fire, with his head upon my knee, like a dog, although he is as tall as a donkey. I hope to get more on the Falkland Islands, where they browse on the poorest land, and their flesh is like venison. The wool is thick, but I fear not like that of the alpaca." In solitary instances, and when taken only a few months' old, the difficulty of taming and domesticating the guanaco, I repeat, unquestionably exists; but it would be quite another thing if we had to deal with a number of them, and at the same time a most difficult task to obtain them as sucklings, and then rear and prepare them for a long voyage. Alonso de Ovalle, in his description of Chili, (Rome, 1646,) speaking of the guanaco, says, "that it is swifter than any horse, but its domestication was never thought of." He adds, "that in his time, flocks of four or five hundred were sometimes pursued by horsemen with dogs, and that this was the most amusing kind of chase which could be wished." While engaged in some important business up the country, and at a considerable distance from the capital one of the Incas received a communication brought to him with more than ordinary speed. To mark his sense of the service rendered, in taking the dispatch from the messenger's hand, (who of course had travelled on foot,)—" *Tia-huanaku*," (Sit down, guanaco,) condescendingly said the monarch to him,—the name which the place, not far from La Paz, where the incident happened, has ever since retained.

As regards public utility, it will have been seen that the great error committed by our early breeders was, that they considered the few alpacas which arrived safe, as mere ornaments to the lawn, never imagining that, by care and attention, they might be converted into profitable stock for farms, situated on ground resembling that upon which they were born. If they had studied the history of Andes sheep, they must, however, have known that these creatures can bear the alternations of cold and moisture without injury, and that at home they are only folded at one particular season of the year—why then shut them up in cribs and cages? Instead of being sent to the hills, they were located on low and rich pastures, in the neighbourhood of towns, feeding upon lands worth £2 per acre, when neglected wastes and commons would have been preferable. Science may be indebted to the several gentlemen who have procured and reared these animals; but, after more than thirty years' acquaintance with the value of the alpaca, it is lamentable to reflect that, in a country like this, only one fair trial of the animal has been made,

---

* The guanaco destined for Josephine Bonaparte, and seen at Buenos Ayres in 1805, was fully as tall as any llama in the flock; but so exceeding fierce and ungovernable that it could not be trusted out of its cage.

REMARKS ON THE SAME—ERRORS COMMITTED BY OUR EARLY BREEDERS.

in Ireland—a trial which, unfortunately, is now only in its third year, and another in Scotland, bearing the date of ten months.

The communications from Messrs Stevenson, Bell, and Stirling, will, however, have been read with more than ordinary interest. Although Mr. Stevenson's location was not left to choice, and he had no previous knowledge of the best mode of treating them, still, "by feeding his Andes sheep in the same way as Highland cattle, he found it answer remarkably well, in so far as their health and growth were concerned." Knowing, as I do, the tricks practised in Peru, from the circumstance of that gentleman's first pair of alpacas not having bred, I should be inclined to think that they were *machurgas*, or, at least, that the female was one. The breeding of the llama, easily and in the regular course, is, however, by him fully established. His loss in a vicuna, six months gone with young, to a llama, as regards natural history, certainly was deeply to be deplored. This example of a successful cross between the largest of the tame, and the smallest of the wild varieties, proves the information obtained by the French *savants*, in Andalusia to have been correct; and had the offspring fortunately been preserved, it might have served as an elucidation to a general law of organic life, in an instance which has not yet come under the special notice of European naturalists.

Mr. Bell's report completely confirms the opinion which I have long entertained, and often expressed—that if alpaca stock is only suitably located and properly treated, flocks may be easily formed. Nature has evidently adapted this animal for a wider geographical range than its native hills; and by placing the pure breed on those of Ireland, and there allowing the animal to fare as it did at home, Mr. Bell and his friends have the merit of being the first among us who gave the stranger a congenial abode. In breeding animals of this kind, it is only the attentive observer of nature who can succeed. Mr. Bell has exercised a sound judgment, and no doubt his experiment will amply repay him for his trouble and anxiety. He already announces an improvement in the weight and quality of the fleece; and from him we learn another encouraging circumstance—viz. that the wool is well set, and the coat not liable to the tick, or the other vermin which often afflict common sheep. These results accompanied by the example of a two years' old fleece, weighing thirty lbs., and the length of the staple being eleven inches, afford the most triumphant answer that possibly could be given to those persons who question the utility of these animals, or doubt the facilities of rearing them.

Their breeding earlier than on their native hills, is also an important discovery. The whole of Mr. Bell's report is indeed of a novel and interesting character; and it is most fervently to be hoped that he will steadily continue his observations on the progress of the little flock confided to his care. His capacious mind and practical habits, as a farmer, must render his communications upon this subject particularly instructive; and should I, at any future period, be favoured with them, I shall take care to do the contents that justice which I feel confident they will deserve—an assurance which I equally address to others who may chance to peruse these pages; for nothing could be more gratifying to my feelings than to see the owners and breeders of alpaca stock communing with each other through me.

I cannot omit offering to Mr. Stirling the just tribute of my grateful acknowledgments for his communication, the value of which, I am sure, the public will duly appreciate. I learned the fact, of a pair of alpacas being in the possession of a gentleman, not named, at Lennoxtown, through an admirable article on the subject, printed in the *Witness*, Edinburgh paper, towards the middle of May, and copied into some London journals. The view therein taken coincided so perfectly with my own opinions —and indeed the whole article was penned with so much warmth of feeling and earnestness of purpose, and at the same time displayed so much accuracy in the details —that I was forcibly struck with it, and have not hesitated to insert it as a note underneath.* I also avail

---

* Since this work was in the printers' hands, I learned incidentally that three of the Kerry alpacas died suddenly—indeed, in one day. My anxious enquiries were immediately addressed to Mr. Bell, from whom, under date of the 4th of July, I received the following reply, which will be read with unfeigned regret by every one capable of appreciating the loss which that gentleman has sustained:—"It is but too true that, since I had the pleasure of writing to you, we have lost three alpacas from the following causes, which I will explain to you as briefly as I can; and really the occurence at the time grieved me so much, that even now I can scarcely trust myself to dilate on the painful subject. One of them, a very fine ewe alpaca, colour black, with white forehead, was in young; and this was her third lamb. For some time previous to giving birth, I suspected all was not right with her, for she rested a great deal, and could not remain on her legs for any length of time. The result was, that she yeaned a dead lamb, which occasioned an inversion of the womb. We had a skilful doctor in attendance; but, unfortunately, he mistook it for a prolapsus only, or simply a bearing down of the womb, no unusual occurrence in many animals when bringing forth their young, and treated it accordingly. The poor animal seemed to be much better after lambing than before, and ate and walked about as if nothing was the matter, but occasionally exposing the womb, which we readily, as we then supposed, replaced properly, and attended her night and day; but, alas! on the 7th day she died, and I cannot express to you the regret, and truly heartfelt sorrow I experienced at the sad event, which I consider was not only a private, but also a great national loss.

"On opening her, I never before beheld a more healthy animal. Liver, lungs, and intestines, completely sound and well; but the womb was partially inverted, occasioning inflammation, the sure cause of death. The dead offspring was a male, of a brown colour, and apparently full grown. It was still-born; but the cause of its death is unknown, as the dam was in no way disturbed, or received any previous hurt that we were aware of. She had a three years' fleece upon her, which, in some degree, may have been detrimental to her. She was a fine animal, and my greatest favourite. We had then only three remaining, viz. two males and one female, and on the same day the other died. The large white male alpaca was missing from the others, and we found him, after a long search, three fields from the rest, quite dead, although only a short time missed from the others. He soon swelled to three times his original size, and on opening him, we found putrefaction had set in as soon as he had died. The cause and symptoms were evidently arising from poison; the lungs were full of coagulated blood, and, with the intestines, dreadfully inflamed. The kidneys were quite soft and soapy, but the liver perfectly sound, and in a healthy state. It is my opinion, he came by his death from having eaten some poisonous weed; as, some days previous, on account of the female before referred to not being able to stand long enough to enable her to feed herself properly, and her having to eat around her when she was lying, I removed them to a different field from their accustomed pasture, to better or longer grass, and here the poisonous weed, I think, had been picked up. Thus it was that two old ones and one young one went, but in a manner that might have happened to any other description of sheep; but the occurrences, melancholy as they were, cannot be used as an argument that the alpaca will not do in this country. On the contrary, it is a fact that the first indication of an animal's decay is generally seen in the liver, especially in the sheep tribe; but no such indication was perceptible here. These cases, therefore, can only be attributed to a combination of adverse circumstances, which it will ever cause me sorrow to think of, as they might probably have been averted, and not to any disease engendered in the animals themselves by their importation into this country. We have only a male and female now; the latter was bred here, and is now in young. I am glad to say, they are both in excellent health."

* "THE ALPACA OF PERU.—We observed in a late number of a periodical, that there was some expectation of introducing the alpaca into this country on a large scale, for the purpose of wool gathering; but an objection was raised, because besides other adverse circumstances, the climate of Great Britain and Ireland was unsuitable. We beg to disabuse the public mind on this sub-

REMARKS ON THE SAME—ERRORS COMMITTED BY OUR EARLY BREEDERS.

myself of this opportunity to thank the editor for his politeness, in transmitting my note to the individual designated, the circumstance to which I am indebted for so prompt and efficient a reply.

Mr. Stirling looks at the prospect of the alpaca being more generally naturalized in Scotland, with the eye of a philosopher and a patriot; and, short as his experience has been, he nevertheless confirms the most essential points required regarding the properties of this animal—viz. its docility, hardihood, spare diet, and the improvement in its fleece to full seventeen lbs.; besides affording practical results on the economy of feeding and folding, which cannot fail to be acceptable to other experimentalists. I cannot, however, refrain from stating, that although Mr. Stirling may have found that, in the absence of other and more congenial food, his alpacas have shown a preference for beans, still I have strong misgivings, as hereafter noticed more in detail, that substances, so hard and dry, in the end will not prove suitable for them.

Winter-feeding, for some weeks in the year, doubtless with us will be required, and all experiments to ascertain which is the best and most economical provender for our *protegés* are of the utmost consequence; but, judging from the fare to which these animals are accustomed at home, and the scanty provision made by the Andes Indian for emergencies, I feel confident that the lighter and more simple the food is the better for them, and that every thing that is not easy of digestion, or apt to overload the alimentary canal, must be avoided. Dried fiorin grass and *tehh*, I should think, preferable to the usual winter fodder; but in carrying out the great experiment, besides the animals themselves, we ought to procure the plants upon which they feed when snow covers the ground. We might easily import the seeds of the *ichu* grass and the *alfalfa*, and there are also several edible roots in Peru highly deserving of our adoption. For the present, useful experiments might be made with fiorin grass, furze, and heather, in a green and dried state.

In expressing his opinion, as seen in the extracts above produced, the Earl of Derby acknowledges the hardihood of the alpaca race, and that they are easily fed, admitting at the same time that they have bred upon his estate, and "that their wool, instead of deteriorating, is rather improved by our climate." His lordship even goes so far as to say that, "as a matter of curiosity, he has no doubt whatever of their success," but announces his apprehension that "they will not answer to form large flocks, like sheep, unless it be solely of females, as the males do not agree well together." His lordship next argues that the disorder, to which he alleges they are subject, will operate as an impediment to their propagation.

The Earl of Derby is the only experimentalist with whom I have communicated by letter, or conversed, upon this subject, who has anticipated difficulties in the formation of alpaca flocks, and to those difficulties his lordship has ventured to give a specific form. This, I must confess, his lordship does with something like hesitation, and as if he relied more upon the opinions of his servants than his own personal observations; but as the same sentiments may be conveyed to others, and, emanating from so high an authority, might create erroneous impressions, however much at variance with the other testimony above adduced, they ought not to be passed over in silence.

To each objection started, I therefore find myself called upon to offer a specific answer; and, in performing this task, I beg to assure his lordship that I am actuated by no feeling of disrespect towards a nobleman of his exalted rank and refined taste—one to whom science is so much indebted for bringing together, from the remotest regions of the globe, and at an enormous expense, rare and valuable specimens of natural history. I cannot, however, divest myself of the apprehension, that similar opinions may also have been uttered by some of his lordship's agents, and, coming from Knowsley, might mislead. At the same time, I avail myself of this opportunity to express my grateful acknowledgments to his lordship for the very frank and condescending manner in which he was pleased to reply to my queries; and I feel confident that his lordship will readily pardon the liberty which I have taken, when he sees that my object is to have the two points referred to thoroughly investigated—a matter of extreme importance to the public in this stage of the question at issue.

By keeping his alpacas in a menagerie, that is, on a confined locality near a house, partly in stalls, and partly in small inclosures, the Earl of Derby necessarily subjects them to restraints to which they are unaccustomed, affecting both their health and temper. The animals being confided to the care of servants, whose attention in the Knowsley collection must be divided among other objects, possibly requiring more solicitude than they do, also precludes the noble owner from the opportunity of specially watching the wants and studying the habits of his Peruvian favourites. His lordship must therefore, in great measure, depend upon others for information; and unfortunately, in establishments like the one alluded to, an owner is frequently made the victim of his keeper's ignorance, carelessness, or impatience.

On all hands it is agreed that the alpaca is docile and tractable, even more so than the llama, and its gregarious habits never were disputed. In former times the Peruvians avowedly had numerous flocks of both animals, and these flocks, we are assured, were of an enormous size, so large that the keepers were obliged to use the *quipus*,* as a register to make out the returns when required. Similar flocks, as before noticed, although on a small scale, are still retained by the tribes dwelling on the declivities of the Andes, where they may be seen browsing on the open ground, and without enclosures;

---

ject, as there can be no doubt as to the climate being suitable—the difficulty not being in the climate—but in the mean time to get these animals imported into this country in a healthy state. They continue very healthy, we are informed, until they reach the Cape of Good Hope, but after being a very little longer at sea (either from being so long at sea, or some defect in physical strength to endure a long sea voyage) they droop and die—at least to the extent of two-thirds. Some method might be adopted whereby they could be put ashore for a time at the Cape, and re-embarked into a succeeding vessel. But be this managed as it may, we have seen a male and female of the alpaca species at Craigbarnet, Lennoxtown, Stirlingshire, which have been there for the last eight months, and have stood the severe winter without injury, and we are assured they are more hardy than our native sheep—require *less* food, and could exist where sheep would die. There seems hardly to be any kind of food they will not eat—they eat turnips, hay, oats, and beans—they are more partial to meadow than to ryegrass hay. These animals are now in the highest order and in the most perfect health—they are jet black, and follow their keeper like a dog, and are very elegant and interesting. The weight of the fleeces of last year was 17 1-2 lbs. Their worthy owner is of opinion, and to use his own words, 'he anticipates,' says he, 'when the navigation between us and South America is diminished as to *length of time*, which steam will most assuredly accomplish, thousands of alpacas will be brought over—our hills will be covered with them, and they will become a source of great wealth to the proprietors and farmers of the Highland districts; for these animals will thrive upon that kind *of coarse bent*, which neither horse, nor cow, nor sheep will look at or touch."'

* A collection of knots, tied on a bunch of threads of various thicknesses, so as to form units, tens, hundreds, &c., and by their colours and combinations serving for all purposes of computation.

REMARKS ON THE SAME—ERRORS COMMITTED BY OUR EARLY BREEDERS.

but the owners take care to keep the varieties separate, a practice by no means followed at the Earl of Derby's, and which may, in some measure, have given rise to the disagreement of which his lordship complains.

It ought not, however, to be concealed, that in porcreation of these animals there is a difficulty, arising out of natural causes, which at a particular moment partly renders them dependent upon the aid of man. This difficulty, and the manner in which it is overcome by the Peruvian Indians, were first noticed by Hernandez, and nothing can be more accurate than his remarks, as exemplified in the practice of the present day. Without assistance sexual intercourse certainly can and does take place, as seen in the wild races, whose structural formation is the same; but in the tame ones it invariably gives rise to confusion. At the rutting season, Andes sheep become restless and lascivious. In Peru this season commences at the close of October, when, as the estimable writer on the climate of Lima remarks, "all nature seems to be in motion; vegetation assumes a new form; earthquakes and volcanic eruptions frequently occur, and the air is filled with an electric fluid. Every production then glows with fresh fire, and by an active stimulus animals are impelled to the propagation and consequent preservation of their own kinds."*

At this period the working llama has a respite, for it would be unsafe to put a burden upon his back, and indeed dangerous for a stranger to thwart his wishes, or control his actions. Both the tame and wild breeds sometimes fight outrageously for their mates, and instances occur of the combat proving fatal to one or both. To this jealous disposition, as noticed by himself, Mr. Stevenson slightly alludes.

By letter General O'Brien informed me, " that in Peru the rutting season commences in the month of November, when the male alpaca throws off his tame and quiet habits, pursuing the females until he separates from the flock one of his own choice. Her he woos with the most ardent demonstrations, and, if she proves coy and runs away, he follows her through the country for miles, and until his importunities have been successful. At this moment, (continues the general,) the flocks of both alpacas and llamas sometimes break up and disperse, running in different directions through the country, and weeks may elapse before the owner is able to collect them in again. Hence, when this season approaches, the Indian shuts up his sheep, separating the male from the female, in pens, purposely constructed in such a manner as to allow of their putting their faces together, and caressing each other a week or a fortnight before the day appointed to bring them out."

The union of the sexes is, to this day, a national festivity among the Peruvians, resembling a sheep-shearing or a harvest-home with us. The villagers assemble dressed in their best; haunches of llama and alpaca are roasted; *cuyes*, or country rabbits, stewed with *——* and *axi*, and copious libations of *chicha*, or beer, go round. On a large farm these merry-makings last a week or ten days, the evenings being spent in dancing or singing *yaravies*, a plaintive melody peculiar to the Peruvians. Young men also play on an instrument made of a cane, and in shape resembling a clarionet, from which they call forth sounds scarcely to be expected from so coarse a piece of mechanism. The extra *corral*, or pen work, is on this occasion principally performed by volunteers, who are amply repaid for their time and labour with the amusements and good cheer provided for them by their host. The Indians of both sexes look for-

* Dr. Unanue. His work was first published at Lima in 1806 and reprinted at Madrid in 1815.

ward to these gay scenes with anxious expectations, and incidents then occur which often afford topics of conversation for a long time afterwards.

Owing to the extremely lascivious disposition of Andes sheep, great care must be observed when the males are admitted to the females. Both by night and by day the shepherd should be vigilant; for besides quarrelling themselves, were two males allowed to compete for the same female, they might trample her to death. Hence every possible precaution ought to be used. Speaking upon this point, General O'Brien further says, " that should the alpaca ever be introduced into this country on a large scale, and as a national benefit, breeders must adopt the Peruvian mode of separating the male from the female, at least a fortnight before the union of the sexes takes place."

The general then goes on to notice the complaints made by Lord Derby's servants of these quarrels and competitions among the males, showing that they did not observe the necessary precautions, or manage the poor animals in a proper manner. What! may I be allowed to ask, have our farmers no difficulty with their sheep at this season? Do they not keep the ewes and rams in separate inclosures, and choose their own time to bring them together? Little more than this is required for alpacas. It is only at one particular season that feuds among the males occur; but certainly the consequences are not so serious as the Earl of Derby imagines, and, the cause being temporary, there can be no great difficulty in guarding against a recurrence.

In some parts of Peru the *llameros* prepare small folds, in which they shut up one of each sex. The male begins his caresses by antic tricks and boundings; the female at first appears shy and moans, while at intervals each spits at the other. After a day or two they become more intimate, when at length the female, with her forelegs bent under her, and resting on her breast, assumes that position in which only she can receive the embraces of her mate; but this is not a forced prostration on her part. It is, on the contrary, the easy and natural posture which she takes when reposing. If she evinces anything like caprice, and difficulties should arise from her repugnance to assume the position required, the keepers place a slip noose, called *pajal*, on the lower part of the fore-legs, when pulling from behind they trip her up, and lighting on her breast, with their assistance she easily receives the act of generation. The state of excitement into which the male has been worked up, is at this moment so great, that he is immediately afterwards turned out separate and left to repose, never being coupled twice in the same day. One, however, suffices for twenty females.

As soon as the pregnancy of the female llama is declared, she is excused from toil; and under similar circumstances the alpaca experiences more than ordinary care. In both, the period of gestation, in Peru, I have been ascertained, is rather shorter than in our clime, where, according to the testimony of Mr. Stevenson, who seems to have paid particular attention to this point, the young one is not dropped till the year has nearly expired. They have one at a birth, and it is very rare indeed that a deviation from this rule occurs. The offspring is, by the Peruvians, weaned at the end of the seventh month; but as the generative powers of these animals are not developed on the Andes slopes so early as with us—a fact now satisfactorily established both by Mr. Bell and Mr. Stevenson—the age of reproduction in their native land does not commence till the third year, at which period the llama is put to labour.

Both varieties live from ten to twelve years, the usual

age of the camel, and carry their teeth well with their age. Before closing this part of the subject, it may be proper to remark, that the same causes which dispose an animal to fatten, also develope the generative powers; and to this circumstance, as Mr. Bell practically argues, doubtless the earlier maturity of Andes sheep on our pastures is to be attributed. It is desirable with us that the lamb should fall when the winter is over, and green food can be had. This advantage we gain in the stranger from the Pacific shores, whose young is usually dropped in the month of April.

CHAPTER VII.

DISEASES TO WHICH THE ALPACA IS LIABLE—TREATMENT—ERRONEOUS METHOD OF PROCURING THIS KIND OF STOCK.

THAT the Earl of Derby's Peruvian flock, if such it can be called, is not in so healthy a condition as might be wished, I was prepared to expect. The locality and mode of treatment could not answer. The lowlands of Lancashire are no more suited for these alpine animals, than the rich and level sward of Oatlands. In the Zoological Gardens, Regent's Park, the nature of the soil, which consists of a thick ungrateful clay, was soon found to be unsuitable for the animals of foreign growth congregated there. Owing to this circumstance, several valuable specimens died; and, in order to preserve the health of the rest, it was deemed necessary to have oak floors raised above the ground, which not being thought sufficient, orders were given for a thick sub-stratum of dry material to be laid under every inclosure in which the animals were kept, and even on the walks. In this case the disadvantages of the locality were compensated by its vicinity to town; but this is not the way in which a breeder of alpacas, with an eventual object in view, will calculate. The farm-yard is usually low and damp, and consequently cannot be better adapted for this class of animals than the menagerie. It is, indeed, surprising that the foot-rot has not been more prevalent among them.

How, then, could it be expected, that the Earl of Derby's alpacas would continue in good health? Besides, from a passage in Mr. Tayleure's letter, it results that some of them were actually diseased before they reached Knowsley. It is not, however, to be supposed, that on its native mountains either the llama or the alpaca is entirely free from disease. The former, when constantly employed in carrying ores, contracts a disorder which makes its appearance on the skin in the form of an eruption, and, if not attended to, assumes a malignant character. This is supposed to be principally owing to the dust, surcharged with arsenical and other deleterious properties, adhering to the skin; which, when the animal is heated, causes an irruption on the surface, succeeded by an incrustation when the poor sufferer is debarred from the opportunity of bathing, and, as a relief, is obliged to have recourse to scratching.

Certain it is, that nothing can be more injurious to health than the ordinary work in the Peruvian mines. Never, indeed, would the Indians employed in those deep recesses be able to pursue their labour, unless they chewed the *coca*, the balsamic and healing virtues of which serve to counteract the poisonous effects of the earthy particles which they inhale. As a proof of the insalubrity of this drudgery, it may be added that the llamas and *machurgas* confined to mine work are seldom serviceable after the fourth year.

Inca Garcilasso tells us of a plague, wearing all the symptoms of a malignant cutaneous disorder, which attacked the tame as well as the wild varieties, and by the Indians was called *carache*, literally meaning the itch. This epidemic occurred towards the year 1544, and the disorder chiefly showed itself under the belly and round the joints, on those parts most divested of hair, and, spreading outrageously, carried off nearly two-thirds of the country sheep, from which period they have never been so numerous as before. It even reached the guanaco and vicuna, but among them was not so destructive, in consequence of their inhabiting a colder region, and not going so much in flocks as the tame breeds.

This, however, was a rare occurrence, occasioned no doubt by the state of the atmosphere, as it extended to the foxes and other wild animals, and one that has never since befallen the country. It has frequently been remarked in Peru, that both the llama and alpaca, when taken down to the lowland towns, and kept there as pets, perspire as soon as the hot weather comes on, and, if neglected, a scurf forms on the skin. In their new character the coat is, of course, carefully preserved, as being ornamental; but if it is shorn off, and the animal bathed in the cool part of the day, before the system has been heated by exercise, or the natural warmth of the climate, the sufferer in a short time invariably recovers. It therefore follows, that the loss of their fleece at the proper season is serviceable to these sheep, and helps to preserve them in good health.

The cooling remedy above pointed out they themselves seek; for when taken down to the heated atmosphere of the plains, should this rash break out, both the llama and alpaca instinctively go in search of a refreshing stream. This Mr. Stevenson noticed in his llamas, erroneously supposing that it was with a view to allay thirst. No alpaca run therefore, if possible, should be without a rivulet; one, indeed, that in some part has a depth equal to three feet but, if more, it ought to be paled off.

Seldom have the few survivors, safely landed on our shores, been properly treated on their arrival. After a voyage of from three to four months, shut up in a crib, and for a great part of the time necessarily eating unwholesome food, it is not to be expected that they can be in good condition. In six cases out of ten they will have contracted the cutaneous disorder above alluded to, which, if it broke out midway over, and has ever since grown upon them, must have affected the blood and disordered the whole system. Sailors do not feel disposed to look after a charge of this kind, while the malady is sure to be aggravated by boisterous weather, seclusion, and the want of exercise. Sick and jaded, the strangers are given in charge to some farrier or empiric, and the consequences, in such cases, Mr. Tayleure acknowledges, by frankly telling us what happened to himself.

Not along ago, half a dozen of these animals, on being disembarked at Liverpool, were put into a cold and cheerless flagged yard, where they contracted a disorder in their limbs, which carried off one half. Frequently, in the last stage of the disease, they have been purchased by dealers at a risk and for a trifle, and through the effects of mercurial frictions, or other strong treatment, have, to all appearance recovered. Soon afterwards, some of the same stock have been seen in a gentleman's park, where, as the cold season approached, it became apparent that the consequences of the pretended cure were worse than those of the original disease.

Although the cases have been few, it is nevertheless a

DISEASES OF THE ALPACA.

fact, that the same tricks have been played off in selling these animals as are so often practised in horse-dealing. The treatment which they have experienced from some owners has been cruel, if not murderous, in the extreme. To the philisophic mind it must be painful to see these mild and inoffensive creatures dragged from their mountain home, and here railed up in a stall, or immured in a dreary stable, by four o'clock in the afternoon, and perhaps left there till ten the next morning, for all depends upon the whim of a servant—often with a wet and filthy bed under them, and not unfrequently eating the offals of a green-grocer's shop; or travelling about in caravans, to be exhibited like wild beasts. Such treatment could not fail to affect the health and spirits of the prisoners, thus bringing on disorders and premature old age.

Proper treatment is not, however, the only point to be taken into consideration. In my own mind I have long been convinced, that the mode of obtaining these animals in Peru was injudicious, and, as regards the ruinous manner in which they are generally brought over, the facts already adduced will speak for themselves. I have even ventured to think that there are better breeds on the Andes slopes than those usually sent to Europe. The first proposition is placed beyond doubt by the incontestable evidence of General O'Brien, who, a few days after visiting Knowsley, wrote to me thus:

"I think that the mode generally used for bringing the alpaca over to this country is defective. For instance, the captains of ships who arrive on the coast of Peru, give an order for two or three pairs, which are brought down from the interior, say fifty leagues' distance. The captain, who cannot be a judge of the animal, is glad to take what he can get, good or bad, as the first cost is only trifling, say from eight to twelve shillings each. He then puts them on board, with some dried clover. The animals are sure to be old ones, as the aborigines are cunning enough to keep the younger stock for themselves, and one half die before the vessel doubles Cape Horn. The others, which the captain brings to England, not unfrequently are old and past bearing, and even live only a short time—but why? Because, I answer, they are placed on some rich and heavy soil, probably in a park, as I have seen them at the Earl of Derby's and other places. They do not there enjoy the high mountain air; they become sickly, and then probably comes on the mange. Their native home is, at least, 10,000 feet above the level of the sea. The highest and most barren mountains in this country would be more congenial to the animal. Although the Earl of Derby, and others here, take particular care of them, yet those gentlemen must pardon me when I say that they are mistaken. I speak from experience; for I have bred some thousands, and also used them as beasts of burden to carry down the ores from my mines."

Nothing can be more just than these remarks. Too liberal an allowance of rich and stimulating food to an animal extremely abstemious, and habituated to live on coarse and light herbage, and that in small quantities, must be injurious; but, above all, if we are to have alpacas, let us begin by placing them in a suitable climate, the more necessary after a long and tormenting voyage.

The Earl of Derby is, I repeat, wrong in supposing that, in its proper element, the alpaca is more liable to the cutaneous disorder to which he alludes, than common sheep are to the rot; nor, when it occurs, can it become contagious, unless it is through neglect. The privations of a long voyage sometimes bring on the scurvy among sailors; and the same causes necessarily, more or less, affect an animal whose structural formation of the stomach is so peculiar as that of Andes sheep. Much depends upon the care of the captain, and the state and duration of proper food. If he is obliged to have recourse to biscuit, the consequences must be fatal. Not long ago ten alpacas were shipped for Liverpool, where all arrived safe, with the exception of one killed by an accident. Lord Ingestre acknowledges that his alpacas, even coming round Cape Horn, "stood the voyage remarkably well," and this success, no doubt, depended upon the proper management of the animals. The best carriers of cattle are the North Americans, from the Eastern States. In most instances, their losses, comparatively speaking, are trifling—a fact which we experienced, some years ago, when engaged in the importation of merino sheep.

Convinced that one half of the failures in rearing Peruvian stock were attributable either to wrong food or over-feeding, I wrote to Alfred Higginson, Esq., surgeon, of Liverpool, to whom, in 1841, I was indebted for an interesting series of remarks on the stomach and intestines of two alpacas dissected by himself. Knowing that his attention had ever since been directed to the same subject, and that subsequent opportunities had presented themselves to him of further examining the digestive organs of several more which died without any ostensible cause, I requested him to favour me with the results of his last operations, which he politely did under date of May 15th, and in these words:—

"Of the three dissections of alpacas dying in this neighbourhood, the last was, perhaps, the most important, and most characteristic of over-feeding, of which there were, however, signs in all. It may be nearly two years since my examination of the last, which died in a pleasant part of the country, a few miles from Liverpool, and where, as I am informed, the animal had the range of a paddock, with several more of its kind, and had sufficient access to water at all times. I found no fat in the interior cavities of the body of this, or the other animals; but on the surface it was rather more abundant in this than in the other two. It was a female, and the state of the bones showed it to be not quite fully grown.

"The viscera of the chest were in a healthy condition; but those of the abdomen drew my attention as being out of order in, perhaps, several respects. The stomach was much gorged with food, hay and oats; the former very imperfectly masticated, and the latter quite whole. Whether their condition varied in the different cavities I cannot say, as the stomach, being wanted for a preparation, was not cut open, but evacuated of its contents through the œsophagus, with much difficulty. Large quantities of half-digested food loaded the intestines; whole oats and hay, in a still fibrous state, being found in the small intestine, and much hard fæcal matter in the large intestine.

"The intestines were pretty extensively adhering to each other by their peritoneal coat, on which a rough deposit of crystalline particles, of great minuteness, but very numerous, had taken place. This deposit having formed most in the parts most dependent after death, made me think that it was probably of *post mortem* occurrence; and I have lately been confirmed in this opinion, by observing the same to have occurred in a dead rabbit. I thought the coats of the bowels weaker in some parts than is natural, for they gave way very easily, chiefly in the small intestines, on attempting to wash out their contents with water. The head was not opened, and the immediate cause of death may, therefore, have had its seat in the brain; but there is no doubt that such a state of repletion with food, would much predispose an animal to fatal disorders. I have not had such opportunity of observing the diseased state of the alpaca's feet, as to give any definite notions on the subject of its ordinary appearance and course."

The preceding results clearly show that the animal dissected, besides having taken improper nourishment, had been over-fed—the mistake committed by the greater part of our early breeders, and the one which, beyond all doubt, gave rise to many deaths. Mr. Edwards confesses that, at the beginning, his alpacas "had a good deal of hard food—oats, beans," &c., besides grass and hay; but when they died so rapidly, he discontinued the practice, and only gave them grass, hay, and vegetables. Notwithstanding Mr. Stirling's success, I here take occasion to repeat, that the experiment of giving beans to animals accustomed to succulent herbage, is, in my opinion, a dangerous one. Their peculiarly framed stomachs are not adapted for dry and hard food, the best proof of which is their habitual abstinence from water. If, at home, they are ever treated with grain, it is maize or millet, in their green, soft, and milky state. A Peruvian would laugh to see us giving them substances which we ourselves could not masticate until they have passed through the millstones. The herbage which they cull on their native hills, is to them meat and drink, and they vary it according to taste and the season. They select it themselves on a wide range, in this respect evincing a strong instinct; and if it is wished that they should prosper, they must be allowed to do the same with us.

There is not, I feel assured any disorder to which Andes sheep are liable, either at home or here, that could prevent them from being successfully bred in our isles. Mr. Tayleure mentions the disease with which his little flock was afflicted; but insinuates that the circumstance was owing to contact with animals imported subsequent to the possession of his first alpacas. Mr. Edwards remarks, that those he had were subject to the scab, and seldom free from it; but at the same time gives us to understand, that this disorder was attributable to the nature of the food of which the strangers partook. The other breeders agree that they have fared well, even in situations by no means eligible; and their earlier maturity with us is an additional proof that the climate agrees with them, and that on our pastures they find kindly herbage. What will the Earl of Derby say to the success which has attended the removal of two of his own alpacas to the Scotch hills, where we are most solemnly assured that they have not had a day's illness since they arrived?

Incidentally, no doubt, they experience the disorder to which the Earl of Derby and Messrs. Tayleure and Edwards advert; and that disorder, like the itch, unquestionably is infectious, (the scab or mange in animals being only another name for it,) but not unless actual contact takes place. Dr. Unanue, speaking of the climate of Lima, remarks that cold and damp, suddenly coming on, are apt to check perspiration, which produces an irritation on the skin, and this, if neglected, ends in an eruption, and finally in the itch; but that, when taken in time, it is easily cured by a cooling medical treatment. That experienced physician thus expressed himself in reference to that cutaneous affection, noticed at a particular season of the year among the inhabitants of the Peruvian valleys; adding, that it is only when the disorder has become virulent, and venereal mixed with it, that mercury is administered.

The same causes produce similar effects on the alpaca. Soon after leaving the sultry coast of Peru, shut up in a crib fastened to the deck, the poor animals are hurried through the variable latitudes of Cape Horn, where heavy gales frequently occur, accompanied by torrents of rain, which necessarily must affect the prisoner. The first visible symptom is, that the animal experiences a nausea or sea-sickness, and abstains from food; in which case it droops, lingers, and dies. If, however, it has the spirits to accept the dry provender offered, sometimes tainted with bilge-water, guano manure, or otherwise affected by the smell of the vessel, it survives in a weak and languid state; but too often contracts the disorder complained of in consequence of the wet and cold currents of air, under the sails, to which it has been exposed, and through neglect and long standing, the eruption assumes a serious character.

On the bleak and unsheltered mountains of Peru, when the season has been variable and the transitions in the weather sudden and severe, the same symptoms sometimes make their appearance, generally as well as partially, and the ravages extend in proportion to the virulence of the disorder which ensues; but this is not owing to any physical defect in the woolly tribe. The cause is, in fact, purely incidental; and when a distemper of this kind befalls his flock, the Indian's first care is to class the sheep, and then shear and bathe them.

Instances are recorded in our own history, of diseases very generally fatal to sheep, the earliest of which is said to have occurred in 1041, when the greater part of both the herds and flocks in the country were destroyed. Even so late as the winter of 1830, the ravages of the rot were so alarming, that, had they prevailed throughout the kingdom in the same proportion as they did in the midland, eastern, and southern counties, the race of sheep would almost have become extinct. In France, scarcely was a merino left after the winter of 1809. In consequence of these visitations, we have not, however, been deterred from increasing and improving our flocks. Nay, those individuals who most contribute to the extension of the best breeds, are ranked among the special benefactors of their country.

We are assured that the great João de Castro, the boldest captain and most skilful navigator of his day—the European who first explored and described the Red Sea—the viceroy of India, and the man who there established Portuguese dominion, brought over to his native land the first orange-tree ever seen in Europe, and "from which originated," says Murphy in his *Travels in Portugal*, "all that valuable fruitage we possess at this day;" adding, "the service which that hero rendered to mankind, by this act alone, entitles him to the gratitude of posterity; and he himself was not so dazzled with the love of military fame, as not to esteem this gift to his country as the greatest of all his actions." And what is it that we do not owe to the first person who introduced the potatoe?

It is this exchange of productions between the new and the old world, that has so much enhanced the value of the discoveries made both by Vasco da Gama and Columbus. How much, for example, are not the South Americans indebted to the Spaniards, who brought to their shores the first horned cattle, horses, and sheep, now so extensively spread over their wide continent? Those enterprising men were not withheld by the dread of either trouble or casualties; and in Peru alone what have been the consequences? Inca Garcilasso informs us, that oxen were there first seen ploughing in the year 1550, when a cow was worth two hundred dollars, and now one may be bought for two. The same authority testifies, that in those days a horse was scarcely ever sold unless through the death of the owner; but if such an occurrence did happen, the price usually rose from 4000 to 5000 dollars, whereas at present a three-years-old may be had for twenty. Merinos were not seen browsing upon the Andes' slopes till the year 1556, when

DISEASES OF THE ALPACA.

they were worth forty dollars each. Such, however, has been the change, that they now sell for four rials per head; and, till lately, the *arroba* (twenty-five pounds) of their wool might have been purchased for the same sum. Had the importers and breeders of these useful races of animals been discouraged by the extra care required, or by the anticipation of those maladies to which each may be occasionally liable in Europe, would this great benefit to the Peruvians ever have been achieved?

The Andes sheep hitherto brought to England having usually been selected on the coast, and in the manner above pointed out, there is but too much reason to apprehend that, in repeated instances, the purchaser was deceived in his bargain, both as regards age and breed. This would be in perfect keeping with the character of the Peruvian Indian. He is habitually wary, easily dissembles, and would rejoice in the opportunity of overreaching a *chapeton*. To him the llama and alpaca are objects of attachment, and he consequently is separated from them with regret. Father Blas Valera calls them "the domestic animals which God providentially bestowed upon the Indians, congenial to their own mild disposition, and so gentle that a child may lead them." Several travellers among the Andes have gone so far as to assert, that a peculiar and visible sympathy exists between the Indian tender and his flock—a remark which may have been suggested by his mild conduct towards them, and the ease with which he restrains any ebullition of temper.

The *machurga* the Peruvian considers a spurious race —the *élève* of the Spaniard; and if, by a little *chaunting*, he could palm one of them upon a foreigner about to leave the country, he would consider that he had done an excellent days' work. Owing to the beautiful proportions of the cross breed, it has happened that the seller of a pair of these mongrels was complimented on his selection, and even extra paid for furnishing what the purchaser, judging only from appearances, considered as a choice specimen. Besides General O'Brien's authority for the indiscriminate manner in which the animals are usually obtained upon the Peruvian coast, I have other authentic testimony to show, that at Arica, for example, when a couple of pairs have been brought in, pursuant to order, even if genuine, they have seldom presented any thing like uniformity either in appearance, size or age, and consequently could not fairly exhibit the true character of the breed. Like the first merino flock brought over from Spain for George III., they are almost always picked out of inferior, if not refuse stock, and, by an intelligent person, would scarcely have been deemed worth the shipping risk.

Certainly there have been some exceptions; but I am moreover inclined to think that the few pure alpacas of a proper age imported, were not of the best breeds. They were chiefly selected in the province of Puno, part of the southern extremity of Peru, and fifty leagues from Arica, the shipping port. This, however, is not the best spot for Andes sheep. The most valuable breeds come from the central provinces; and here it may not be irrelevant to observe that there are two kinds of alpacas, differing in size, figure, and fleece. The one called *coyas** is the most diminutive, and esteemed for its smallness of bone and symmetry of form.

This breed is chiefly confined to the Cusco range of mountains, more particularly to that part of it intervening

---

* This was the title, equivalent to Highness, given to the Incas' daughters. Tradition says that the name has been retained to the Cusco breed of alpacas, in consequence of select specimens of them having in former times been kept in the palace gardens.

---

between the ancient city of the Incas and Huamanga. It is probably a remnant of the old royal flocks, or those once owned by the priests of the sun; who, as previously stated, had the choicest breeds. That territory was besides the principal theatre of the agricultural operations carried on for account of the Incas, the seat of power, and the centre of Peruvian civilization. Mr. Cross's specimen was of this breed, and one so perfect has not since been seen on this side of the Atlantic. The southern breed is taller, thickset, the wool longer and more shaggy, and the usual colours russety brown and black, the proportion of white being small. Among the *coyas* is found the chocolate brown, remarkable in Mr. Cross's alpaca, the hinder half of which was that colour, and the fore part milk white. Of a small southern flock there is a picture in the Polytechnic Institution, Regent Street, painted at Liverpool by Ansdell.

Imperfect as the experiments made within the British Isles have been, and, with the exception of two, undertaken with the least probability of success, still it is satisfactory to find it incontestably proved that alpacas, on our soil, may be made to breed and thrive; nay more, that the fleece becomes finer and heavier, besides the animal attaining its maturity a year sooner, than in Peru. More we could not desire or expect; and the means therefore of forming them into flocks must depend upon ourselves. The more the subject is considered, the clearer will it appear that the naturalization of this animal is a matter of practical utility—an expedient that may be adopted at a trifling expense, and with general benefit to the community. If, however, we are to have this acquisition for the reasons already alleged, greater care must be taken in the choice and shipment. Above all, they must be genuine; and, when once landed, let us keep them pure and unmixed; for of what use can it be to breed a mongrel race, unserviceable as stock, and in fleece deteriorated?

The object of the flock-master is to raise that breed of sheep which will best pay for its food, and in the shortest time attain maturity, as well as the greatest weight in fleece or carcass. These advantages we at once secure by possessing alpaca stock; but still the trial would be an unfair one, unless the animals are placed on a suitable locality. Writing to me upon this subject, General O'Brien expresses himself thus:—

"Since my arrival here, (Liverpool,) I have seen some very beautiful articles manufactured from alpaca wool, which astonished me: and I then thought of paying a visit to the mountains of Wicklow and Kerry to ascertain their height and situation. This I did; and on investigation found that they would be admirably well adapted for the growth of those animals; more especially the hilly lands on Lord Wicklow's estate, and part of that belonging to Earl Fitzwilliam. I have not been in the Highlands of Scotland, but I dare say that alpacas would equally thrive there. I can venture to affirm that, if the animals are only properly chosen, and then prepared some little time previous to their embarkation for this country, they would arrive in good health, and, eventually become a source of great national wealth, besides promoting the interests of a certain class of our manufacturers. I have lived among the animals twenty years, and consequently ought to have some knowledge of their habits and the treatment which they require. I should have been delighted to have devoted some years to the cultivation of their wool; but I am now going abroad,* and could not have the same interest in the

---

* Soon after closing his correspondence with me, the General embarked for Monte Video, in order to superintend a pastoral establishment formed on the eastern banks of the La Plata.

pursuit as other persons who remain at home, and whose mountain lands at present produce nothing, but which by the introduction of alpacas might be made as profitable as the very best."

CHAPTER VI.

HOW THE BEST BREEDS MAY BE OBTAINED.—SAFE AND ECONOMICAL MODE OF BRINGING THEM OVER.

Once admitted that the applicability of the alpaca to our soil and circumstances has been sufficiently tested, and that the possession of flocks of these animals is an object of great and national interest, all we require, in order to realize the ends proposed, is a sufficiency of genuine stock, at a price so reasonable as to bring it within the reach of the second-class farmer. As regards the selection, General O'Brien recommended that two English shepherds should be sent to Peru, for the purpose of collecting the animals and taking care of them on their passage over. I entertain a totally different opinion, and this is perhaps the only point on which I differed with that experienced gentleman, at the time we carried on our correspondence upon this interesting topic. I should rather suggest the expediency of sending over an individual acquainted with the manners, institutions, and language of the country—one likely to meet with the consideration of the ruling authorities and the sympathies of the inhabitants; for although there could be no objection to the occasional embarkation of a few pairs of these, to the Peruvians, truly national animals, there might at first be some difficulty in obtaining permission to export a large number. Some small duties might also be exacted; and, as it would be advisable to avoid all clandestine operations, negotiations might be requisite which could not devolve upon an ordinary person.

The selection of the best breeds in the interior, and the conveyance of them to the coast, are alike matters of importance. More minute information regarding the habits and treatment of the animals, in health and in sickness, would also be desirable. These are duties, the performance of which is beyond the capacity of a shepherd, and who besides, by being dependent upon strangers in all he had to do, would constantly be deceived. A man of address, intelligence, and observation, ought therefore to be preferred—one who has some reputation at stake, and who from patriotic motives would feel disposed to devote his attention exclusively to the undertaking. Besides being deeply impressed with the magnitude of the object in view, he should also have some preliminary acquaintance with the subject, and be prepared to meet contingencies.

To such a person as this the selection ought to be confided. This duty cannot be properly performed by merchants established upon the coast, although they may be the most eligible to conduct the shipments; and it ought invariably to be borne in mind, that it is on the choice of the breed that the success of the scheme proposed mainly depends. I should rather advise that each shipment, if large, be accompanied by a Peruvian *llamero*, or shepherd, one accustomed to manage these animals, acquainted with their tempers, and experienced in the cure of their diseases. Young men of this class might easily be had at a trifling expense; and, if Indians, a little tuition and intercourse with Europeans would change their disposition, and induce them to improve their habits. They would be the most proper persons to feed and nurse the animals on the voyage; and, on their arrival, they might besides serve as valuable instructors to our people. Mr. Cross has always found that natives best understood the management of his giraffes, and in that of alpacas scarcely can it be alleged that we have had practical experience; but, above all, the keeper's attention should not be directed to other objects. His whole time ought to be devoted to the charge confided to his care.

Before I close this part of my subject, it may be proper to remark, that the alpacas should be purchased as soon after they are weaned as is practicable, and placed in a depot, for which purpose General O'Brien assured me that a suitable estate might easily be rented on the coast, and at a very cheap rate. Here it would be well to keep them in a kind of preparatory school, and partly on dry diet, at the same time that all exceptionable members of the flock should be withdrawn. It would also be advisable to have them classed according to their ages, and ship them when a year and a half old. On being landed upon our shores, they ought immediately to be conveyed to the hills, instead of being confined in some suburban inclosure, however short the period may be. Let them at once know their homes, and, freed from restraints, let them enjoy the society of their own species. On reaching their destination, they should, moreover, be carefully inspected, and if an eruption appears upon the skin of any, they should be separated from the rest, and placed in a depot by themselves. This would be the best way to treat the poor animals, and by this means also we should provide for a healthy progeny. Every time they are shorn, they ought also to be carefully examined.

If possible, the boisterous, tedious, and ruinous voyage round Cape Horn ought to be avoided; and fortunately there is every likelihood that a safe and practicable expedient will soon present itself. When the Pacific Steam Navigation Company was formed in 1840, it was confidently expected that a regular line of communication would be established from Chili, along the Peruvian coast, to Panama—an undertaking rendered the more easy and sure by the known existence of coal upon several contiguous points. Through mismanagement and misunderstandings, the hopes of the public were however, in this respect, disappointed, two steamers only having been put on; but a new company has now been organized by the original proprietor, Mr. Wheelwright, for the purpose of navigating the Pacific from Peru to Panama, and there is every certainty of this great desideratum being realized in the early part of next year.

The plan proposed is to put on a requisite number of iron steamers, of a competent size, with accommodation for 100 passengers, and stowage for 250 tons of merchandise. The sea being perfectly smooth, the animals might be placed upon the decks of these vessels, and in from five to seven days reach the isthmus. Landed there, they might easily travel over the hills to Cruces, a distance only of five leagues, where, until a suitable road has been opened across, they can be embarked in the flat-bottomed boats used in the country, and, with the aid of a strong current, always descending, pass down the river Chagre to the port of the same name, situated on the Atlantic, in from ten to fifteen hours.

The little towns of Cruces and Gorgona stand upon elevated parts of the banks, and are healthy. Small farms are also seen along the water line to the sea, chiefly cultivated by free blacks, where, by proper arrangements and timely orders, any quantity of fodder and esculent roots might be provided. The inhabitants are a kind and hospitable race, and would readily be induced to second the efforts of the tenders to bring the flocks across. If necessary, temporary depots might be formed on the banks of the river, where, for a few days, the travellers would find rest and refreshment; and, should any of them have suffered from the previous trip, they might be left behind till perfectly recovered. The dry season commences in December, and terminates in June or July. In that interval, the operations in question ought therefore to be carried on.*

From the port of Chagre, the alpacas might be conveyed to England in suitable vessels, and at an easy freight. The gun-brigs sold by the Admiralty, as well as the fast-sailing vessels thrown off the West India packet line through the introduction of steamers, would answer admirably well for this purpose, and might be fitted up at a small charge. The animals would thus reach their destination in six or seven weeks from the period of their embarkation in Peru, with the advantage of a short rest and a fresh supply of food; thus avoiding the dangers and delay of the usual route, the circumstance to which so great a mortality among them is to be ascribed.†

By means of a practical and well-directed effort, similar to the one here pointed out, it will appear evident that genuine alpaca stock may be obtained quickly, and at a moderate price; and as experience—that great teacher in all new undertakings—has so far overcome the difficulties opposed to the naturalization scheme, as to show the nature and extent of the errors committed by our early breeders and amateurs, when we once receive the pure breed, and are satisfied on the score of age, we now know how to locate and treat them. All the necessary elements of success in fact exist—are within our reach, and nothing more is required to render them available than a few appropriate combinations. The active co-operation of a small number of capitalists and land-owners alone would suffice; and that this great measure of improvement may be carried into effect at much less risk and expense than is generally imagined, will appear from the subjoined estimates.

In consequence of the abundance of dried grass, principally *alfalfa*,‡ a species of lucerne, and esculent roots, alpacas on the Peruvian coast may be shipped and provisioned to Panama at £1 per head. Agreeably to the plan suggested, let an adequate stock for the British Isles be set down at 10,000 females and 500 males, clear, with 1050 extra ones in the same proportion, to make up for any deficiency occasioned by deaths on the passage over, and calculated at about ten per cent. The prime cost and conveyance to England, in that case, would stand thus:—

11,550 alpacas shipped in Peru at £1 each,	£11,550
Freight to Panama, at £1 each,	11,550
Conveyance from Panama to Chagre, 10s. each,	5,755
Freight to England at £3 each,	34,650
Contingencies, rent of estate, agents, &c.	6,000
Prime cost and conveyance to England,	£69,505

* An instructive article on "the best means of establishing a commercial intercourse between the Atlantic and Pacific oceans," will be found in *Blackwood's Edinburgh Magazine* for November, 1843.
† In 1800, Josephine Bonaparte added to her collection at Malmaison a pair of llamas, sent overland from Peru to Carthagena. They were attended by an Indian, and bore the journey well. From Carthagena they went to St. Domingo, had a little rest, and thence were conveyed to France, where they arrived in good health, and lived many years.
‡ On the level spots, even near the coast, the inhabitants sow the seeds of this plant, and afterwards water the grounds by

CHAPTER IX.

NATIONAL ADVANTAGES WHICH WOULD ACCRUE FROM THE NATURALIZATION OF THE ALPACA.

If sold, on being landed, at £12 each, and the usual price has been from £20 to £30, the surviving 10,500 alpacas would realize £126,000, thus leaving a profit of £56,495; but if placed on farms, and allowed to breed, at the end of sixteen years the results would approximately be as follow, calculating that the animals cease bearing at the end of twelve years:—

	Females.	Males.
1st year, 10,000 females would produce	5000	5000
2d do. do. do. do.	5000	5000
3d do. do. do. do.	5000	5000
4th do. do. do. and 1st female brood.	7500	7500
5th do. do. do. and 1st and 2d female broods,	10,000	10,000
6th do. do. do. and 1, 2, and 3d do.	12,500	12,500
7th do. do. do. and 1, 2, 3, and 4th do.	16,250	16,250
8th do. do. and 1, 2, 3, 4, and 5th do.	21,250	21,250
9th do. do. and 1, 2, 3, 4, 5, and 6th do.	27,500	27,500
10th do. do. 1, 2, 3, 4, 5, 6, and 7th do.	30,625	30,625
11th do. do. 1, 2, 3, 4, 5, 6, 7, and 8th do.	41,250	41,250
12th do. do. 1, 2, 3, 4, 5, 6, 7, 8, and 9th do.	55,000	55,000
13th do. do. 1, 2, 3, 4, 5, 6, 7, 8, 9, and 10th do.	70,312	70,312
14th do. 1, 2, 3, 4, 5, 6, 7, 8, 9, 10, and 11th do.	88,437	88,437
15th do. 3, 4, 5, 6, 7, 8, 9, 10, 11, and 12th do	113,437	113,437
16th do. 4, 5, 6, 7, 8, 9, 10, 11, 12, and 13th do.	146,093	146,093
Total Produce,	633,154	633,154
Add Males,	633,154	
	1,266,308	
Deduct 20 per cent for sterility and deaths,	253,261	
Net Produce,	1,013,047	

From the above period the increase becomes extremely rapid, and the advantages consequently proportionate; but supposing that, at the end of the sixteenth year, it should be determined to wind up the speculation by disposing of the stock on hand, in that case 1,013,047 alpacas, at £5 each, would realize £5,065,235, an enormous sum when the original outlay, and the cheap rate at which they can be fed and tended, are taken into account. The valuation of £5 per head will not be deemed excessive, if the prospective profits on this kind of stock are duly considered, jointly with the high prices at which common sheep, of a choice breed, have always sold among us.

As before stated, each alpaca annually yields a fleece, weighing, on the most moderate calculation, 10 lbs. of clean wool, which, sold only at 1s. 6d. per lb., would realize 15s., a much better remuneration than can be obtained from any other mode of farming, with the prospect of an advance to 17 lbs. per fleece, the acknowledged clip of several already shorn in the kingdom, besides that of a higher value for the wool. It is, moreover, to be borne in mind that the carcass, the usual weight of which is 180 lbs., at 6d. per lb., would sell for £4 15s., and the skin may be safely set down at 5s. more. It is further to be recollected that a year will be gained in the produce of the flock, the preceding estimates of increase being made out as corresponding to Peru.

There is still another part of the speculation which must not be overlooked. For sixteen years the parties

means of *acequias*, or canals of irrigation, whereby an abundant crop is obtained. Along the sides of these fields, where the water is most plentiful, rows of willows are planted, the tender twigs of which are eaten by cattle, and the rest used for basket work.

NATIONAL ADVANTAGES OF NATURALIZATION.

interested will have been in the receipt of an annual income, arising out of the fleeces produced by the original stock and successive broods, male and female, in something like the subjoined proportions:—

				Fleeces.
1st year, 10,500 stock would yield				10,500
2d do.	do.	do.	do.	10,500
3d do.	do.	do.	do. and 1st brood,	20,500
4th do.	do.	do.	do. and 1st and 2d do.	30,500
5th do.	do.	do.	do. and 1, 2, and 3d do.	40,500
6th do.	do.	do.	do. and 1, 2, 3, and 4th do.	55,500
7th do.	do.	do.	do. and 1, 2, 3, 4, and 5th do.	76,000
8th do.	do.	do.	do. and 1, 2, 3, 4, 5, and 6th do.	101,000
9th do.	do.	do.	do. and 1, 2, 3, 4, 5, 6, and 7th do.	133,000
10th do.	do.	do.	do. and 1, 2, 3, 4, 5, 6, 7, and 8th do.	145,000
11th do.	do. and 1, 2, 3, 4, 5, 6, 7, 8, and 9th do.			200,000
12th do.	do. 1, 2, 3, 4, 5, 6, 7, 8, 9, and 10th do.			261,350
13th do.	do. 1, 2, 3, 4, 5, 6, 7, 8, 9, 10, and 11th do.			343,850
14th do.	do. 2, 3, 4, 5, 6, 7, 8, 9, 10, 11, and 12th do.			443,350
15th do.	do. 3, 4, 5, 6, 7, 8, 9, 10, 11, 12, and 13th do.			573,474
16th do.	4, 5, 6, 7, 8, 9, 10, 11, 12, 13, and 14th do.			729,848
Total fleeces,				3,174,872
Deduct 20 per cent for sterility and deaths,				634,974
Net Fleeces,				2,539,898

2,539,898 net fleeces, at 10 lbs. each, make 25,398,980 lbs. of wool, which sold only at 1s. 6d. per lb., would realize L.1,904,923, an enormous sum, applicable to the reimbursement of the first outlay of L.69,505, and the payment of the current expenses at home. The proceeds of the increased progeny might, in fact, almost be set down as so much clear gain. At first sight these results, produced by so small a capital, may appear excessive, being out of all proportion to the profits usually made by our farmers; and it may be argued that a depreciation in the price of the animals and their wool is likely to follow, in consequence of the great increase of both.

It is, however, to be borne in mind, that the preceding estimates are founded upon positive data, and large allowances have been made for casualties. As regards a decline in the value of the stock, it might safely be contended that the probability is rather in favour of an advance; for, if breeding alpacas once becomes fashionable, and in no country has example so powerful an influence as in ours, the home demand necessarily must rise. As for the wool, that of common sheep must fall very low indeed, when alpaca will not sell for 1s. 6d. per lb. The Liverpool trade report of May 23, states thus:—"There has been an animated enquiry for alpaca wools this week, particularly for black and brown, and the sales made have been at fully previous prices, say from 1s. 8d. to 2s. 6d."

But, to pursue this part of my subject—if, at the expiration of the sixteen years, we should have any surplus stock to dispose of, there is every chance of its meeting a good market in the north of Europe; the more so as, including casualties, alpacas cannot be imported direct from Peru at a less charge than from L.8 to L.9 per head; and those born and bred upon our soil would be infinitely preferable. Should any thing like competition arise, there is, therefore, no likelihood whatever of the value of these animals falling below the cost of their importation.

Alpaca wool being suited only for the finer class of goods, and calculated to compete in some degree with silk, no very great depreciation in its value can be looked for, unless a complete revolution takes place in our woollen manufactures, and in the raw material with which they are fed. The united flocks in Great Britain are estimated at thirty-two millions of sheep, averaging 4 lbs. of wool each. Our yearly clip may therefore be set down at 128,000,000 of lbs., which through an acknowledged deterioration in the quality, owing to the weight of carcass being consulted rather than the fineness of the fleece, advantages which experience has proved to be incompatible, to the grower would not realize more than 6d. per lb., or total L.3,200,000.

Should we, agreeably to the suggestions here held out, provide ourselves with alpaca stock to the extent of 1,013,047 head, which, with a very small sacrifice, I repeat, may easily be accomplished, and within the term above specified, we thereby annually create a new capital, equal to L.759,786, represented by 10,130,470 lbs. of wool, rating the fleece at no more than 10 lbs., and the price 1s. 6d. per lb. Calculating only that the raw material is tripled in value by the application of skill and labour, it follows that the new capital created would be raised to L.2,273,358, an amount which our countrymen might every year continue to earn, with the prospect of a rapid advance.

It is believed that our woollen manufactures, to feed which we are obliged annually to import an immense amount of foreign wool, principally from Germany, and at an extravagant price,* afford employment to upwards of 500,000 persons, independent of the growers of that proportion furnished by ourselves. The annual value of this class of manufactures is estimated at from eighteen to twenty millions of pounds sterling, from which, calculating the average rate of tax on expenditure at 16 per cent, the government derives a revenue of at least L.3,000,000. Without in the slightest degree interfering with those establishments already formed for the purpose of supporting this immense and valuable branch of national industry, alpaca manufactures, when we once possess a sufficiency of the raw material at home, might easily be raised in such a manner as to become equivalent in value to one-fourth of our other woollens, thus affording employment to a corresponding number of hands, and to the government a proportionate revenue.

Since the year 1825, when the restrictions on the shipment of home-grown wools were removed, a new branch of commerce opened upon the country. In 1833, we exported 4,992,110 lbs. of British raw wool, principally to Belgium, France, and the United States, which in 1837 declined to 2,831,352 lbs., and in 1838, to 1,897,360 lbs.; but in 1840 again rose to 4,810,387 lbs.; in 1842, to 8,578,691 lbs., besides 5,962,191 lbs. of woollen yarn; and in 1843 to 8,179,639 lbs., independent of 7,410,313 lbs. of the same kind of yarn. The exportation of British raw wools and woollen yarn is consequently on the increase, although it must be taken into account that a proportion of the latter was spun from alpaca, upon which the spinners made a large profit. Once provided with stock, which would place us more on a level with the French, who grow their own silk, we might equally export the raw material as we now do the yarn, which would besides help to vary the commodities supplied to the foreign market.

By means of the very simple operation above explained, we therefore have it in our power to render an essential service to the country at large, and more especially to befriend a considerable number of our distressed artizans. Hitherto we have been entirely dependent upon a foreign and distant country for a raw

---

* In 1842, the quantity of sheep and lambs' wool imported into the United Kingdom was 44,463,542 lbs., of which 417,401 lbs. were exported, leaving the home consumption at 44,022,141 lbs. In 1843 we imported 47,335,559 lbs., exported 892,867 lbs., and consumed 46,443,052 lbs. Of this supply 15,613,569 lbs. in 1842, came from Germany, and in 1843, the quantity from the same quarter advanced to 16,835,114 lbs.

NATIONAL ADVANTAGES OF NATURALIZATION.

material, now very successfully introduced into our factories, and rising in demand as people acquire the taste and power to consume it—a raw material which, it is admitted, we can easily grow even under an improved form, and without throwing the smallest impediment in the way of any of our other agricultural pursuits; and when grown, spin and weave it, without in the slightest degree abridging the labour expended upon any one branch of manufactures already established.

Every year's experience demonstrates the certainty of an increased demand for alpaca stuffs and yarn, to keep pace with which an additional quantity of wool will be called for. The shipments in Peru, as before pointed out, have been carried to their full extent; and the supply for the last three years proved to be so scanty that the mixture of other wools is adopted, to the detriment of the genuine texture. A foreign supply is besides always precarious; nor ought it ever to be forgotten that, in 1809, through the incidents of the war, Spanish wools, which with us at that time held the place those from Germany now do, rose as high as 8s. and 9s. per lb. Within five years, alpaca wool alone has created a trade worth nearly half a million sterling per annum, and it will be our own fault if that sum is not quadrupled in double the same period.

The productiveness of our soil, I repeat, is far from having attained its limit; and, with prospects of a successful issue before us, that zeal for the advancement of agriculture—that anxiety for the employment of our suffering population—so widely and so loudly manifested at this particular juncture by the higher orders in society, can never be held to be sincere, if the plan here suggested, of turning our waste lands to account, and extending a branch of manufacture, commenced under the best auspices, is any longer neglected.

CONTENTS.

		Page
CHAPTER	I.—History and Properties of the Alpaca,	1
"	II.—Alpaca Wool and Meat,	6
"	III.—Applicability of the Alpaca to our Soil and Circumstances,	8
"	IV.—Benefits which would accrue to the British Farmer and Manufacturer from its Naturalization,	10
"	V.—Results of the Experiments already made to Naturalize the Alpaca,	14
"	VI.—Remarks on the same—Errors committed by our early Breeders,	20
"	VII.—Diseases to which the Alpaca is liable—Treatment—Erroneous method of treating this kind of Stock,	24
"	VIII.—How the best Breeds may be obtained—Safe and economical mode of bringing them over,	28
"	IX.—National advantages which would accrue from the Naturalization of the Alpaca,	29

www.ingramcontent.com/pod-product-compliance
Lightning Source LLC
Chambersburg PA
CBHW062207220526
45470CB00009B/2953